Electric Aviation

Electric Aviation

Harper Brooks
Jason Page
Claire Russell

Kruger Brentt
Publishers
2026

Kruger Brentt Publishers UK. LTD.
Company Number 9728962

Regd. Office: 68 St Margarets Road, Edgware, Middlesex HA8 9UU

ISBN: 978-1-78715-656-2

For information on all our publications visit our website at http://krugerbrentt.com/

Preface

Electric Aviation explores one of the most transformative frontiers in modern transportation—the shift from conventional fossil-fuel-powered aircraft to cleaner, quieter, and more efficient electrically propelled flight systems. As climate concerns intensify and global air travel continues to expand, the aviation industry faces unprecedented pressure to reduce emissions, noise pollution, and dependency on nonrenewable energy sources. This book presents a comprehensive introduction to the science, engineering, and technological innovations driving the electric aviation revolution.

The text begins with the foundational principles of electric propulsion, motor design, energy storage systems, and aerodynamics, laying the groundwork for understanding the engineering considerations behind electric and hybrid-electric aircraft. It then moves into contemporary advancements such as next-generation batteries, energy-dense materials, solar-assisted propulsion, hydrogen-electric concepts, and distributed electric propulsion architectures. These developments, once merely conceptual, are now rapidly gaining traction as governments, industries, and research institutions invest heavily in sustainable aviation technologies.

Beyond the technical aspects, the book also investigates the broader ecosystem surrounding electric aviation, including regulatory frameworks, safety standards, airport infrastructure, environmental impact assessments, and economic feasibility. It critically examines the challenges associated with battery limitations, thermal management, grid integration, and long-distance flight, while highlighting emerging solutions and areas of active research.

Whether for students, engineers, aviation professionals, policymakers, or sustainability advocates, this book provides a clear, accessible, and forward-looking resource on the technologies redefining the future of flight. By combining scientific rigor with real-world applications and case studies, it aims to inspire innovation and contribute to global efforts toward greener and more sustainable aviation systems.

Harper Brooks

Jason Page

Claire Russell

Contents

CHAPTER 1

Introduction to Electric Aviation

1.1 DEFINITION AND SCOPE OF ELECTRIC AVIATION

Electric aviation is defined by the shift toward electrified propulsion systems, encompassing battery-electric, hybrid, and hydrogen fuel cell technologies. Its current scope is firmly rooted in short-range, light-payload applications such as flight training, urban air mobility, and nascent regional transport. While significant technological and regulatory challenges currently limit its application to larger, longer-range commercial aircraft, the continued maturation of energy storage and power systems promises to progressively widen this scope. Electric aviation is not merely a futuristic concept but an actively evolving sector that has already begun to alter the trajectory of aerospace development, aiming for a cleaner, quieter, and more connected future.

1.1.1 Defining Electric Aviation

At its core, electric aviation is defined by the use of electric motors for propulsion, either wholly or partially, reducing or eliminating the reliance on internal combustion engines or gas turbines. Unlike traditional aircraft that burn kerosene or avgas to generate thrust directly or to turn a propeller, electric aircraft utilize electrical energy to power motors that drive fans or propellers. This electrical energy can be sourced in various ways, leading to distinct sub-categories within the definition. The most purely defined form is "all-electric" or "battery-electric" aviation, where the sole source of onboard energy is a battery pack. These aircraft produce zero direct emissions during flight and are mechanically simpler than their combustion counterparts.

However, the definition also extends to "hybrid-electric" systems. Much like hybrid automobiles, these aircraft combine conventional fuel-burning engines with electric propulsion systems. In these configurations, the combustion engine might drive a generator to keep batteries charged or provide supplemental thrust during high-demand phases of flight, such as takeoff. Another emerging category that falls under the broad umbrella of electric aviation is hydrogen-electric propulsion. In these systems, hydrogen fuel cells generate electricity through a chemical reaction, which then powers the electric motors, with water vapor as the only emission. Therefore, electric aviation is best understood not as a single technology, but as a spectrum of electrified propulsion architectures designed to increase efficiency and reduce environmental impact.

1.1.2 Core Technologies Driving the Revolution

The scope of electric aviation is intrinsically linked to the maturity of its underlying technologies. The primary driver—and currently the primary limiter—is energy storage. Electric motors themselves are highly mature, offering remarkable efficiency (often exceeding 90%) and high power-to-weight ratios.

The challenge lies in storing enough potential energy to power them for useful durations. Jet fuel is incredibly energy-dense, containing roughly 50 times more energy per kilogram than current state-of-the-art lithium-ion batteries. While electric motors use energy far more efficiently than combustion engines, this massive disparity in energy density means that all-electric aircraft are currently restricted to shorter ranges and lighter payloads.

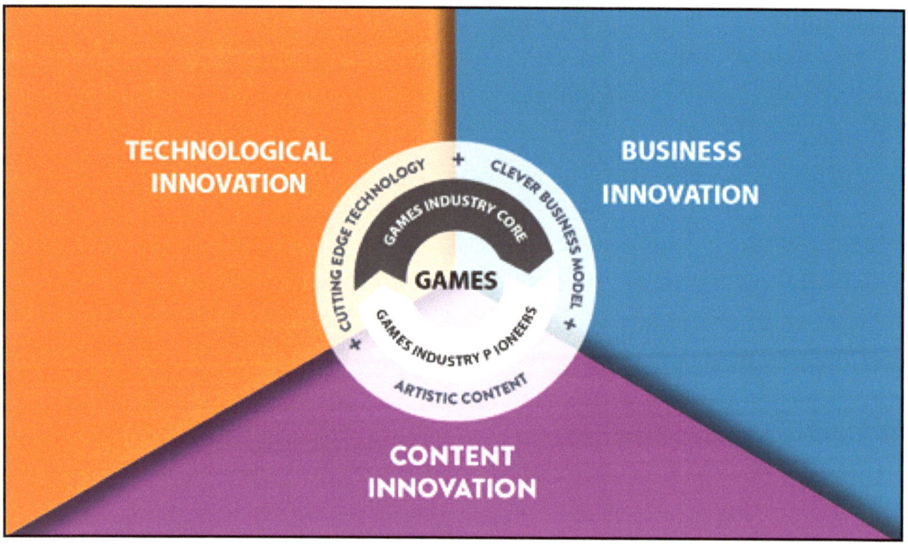

Fig: Games are the Driving Force of the Digital Revolution

Beyond batteries, developments in power electronics are crucial. These systems manage the massive flows of high-voltage electricity between batteries, generators, and motors. They must be incredibly reliable, lightweight, and capable of handling the immense thermal loads generated during operation. Furthermore, advances in lightweight composite materials are essential to offset the heavy weight of battery packs, allowing electric aircraft to achieve viable payload capacities. The convergence of these technologies defines the current practical boundaries of electric flight.

1.1.3 Current Scope and Near-Term Applications

Despite the technological hurdles, the scope of electric aviation is already expanding into tangible, commercial applications. The industry is not waiting for a "miracle battery" to begin implementation; instead, it is finding niches where current technology offers immediate benefits. The initial scope is focused on smaller vehicles and shorter distances, where the limitations of battery energy density are less prohibitive. This has led to a burgeoning sector of diverse aircraft designed for specific mission profiles that were previously impractical or too costly with conventional engines. Current and near-term applications are actively reshaping the following segments:

- **Urban Air Mobility (UAM) and eVTOLs**: Perhaps the most hyped sector, Electric Vertical Takeoff and Landing (eVTOL) aircraft are designed for short intra-city hops, bypassing ground traffic. Dozens of companies are currently flight-testing these vehicles, aiming to launch "air taxi" services within this decade.
- **General Aviation and Flight Training:** Small, two-seater all-electric aircraft are already certified and in use for pilot training. Their short endurance (typically around one hour) is perfectly suited for training circuits, offering massively reduced operating costs and noise levels for flight schools.
- **Unmanned Cargo Drones:** Before carrying passengers, many electric platforms are being deployed for logistics. Heavy-lift drones are expanding the scope of middle-mile delivery, capable of moving essential supplies to remote areas without the infrastructure required for traditional aircraft.
- **Regional Short-Haul Commuters:** Several companies are working on 9 to 19-seat electric or hybrid aircraft designed for short regional routes (e.g., 100-250 miles). These aim to reconnect smaller communities that lost air service due to the poor economics of operating large turboprops on thin routes.
- **Retrofit Programs:** Rather than building entirely new airframes, a significant portion of the current scope involves retrofitting existing, proven airframes (like the Cessna Caravan or De Havilland Beaver) with electric powertrains to accelerate regulatory approval and market entry.

1.1.4 Challenges Limiting the Scope

While the potential is immense, a realistic assessment of the scope must address the substantial barriers that currently constrain the industry. The transition to electric propulsion is arguably harder than the transition from piston engines to jet engines in the mid-20th century because it requires a fundamental change in the energy source itself, not just the mechanism of converting it to thrust. These challenges determine the pace at which the scope of electric aviation can expand from niche applications to mainstream adoption. The most critical limiting factors include:

- **Battery Energy Density:** Current lithium-ion technology is too heavy for long-haul flights. Significant breakthroughs in chemistry, such as solid-state or lithium-sulfur batteries, are required to make larger all-electric aircraft commercially viable beyond short hops.
- **Certification and Regulation:** Aviation regulatory bodies like the FAA and EASA maintain incredibly high safety standards. Certifying novel propulsion systems, high-voltage architectures, and new autonomous flight software is a slow, complex, and expensive process without established precedents.
- **Infrastructure Requirements:** Widespread adoption requires entirely new ground infrastructure, including high-capacity charging stations at airports and vertiports, which must be integrated into already strained municipal power grids.
- **Thermal Management:** Electric powertrains generate significant heat that must be dissipated. Managing these thermal loads in the confined, lightweight structure of an aircraft, especially during high-power phases like takeoff, presents a complex engineering challenge.
- **Economic Viability and Scaling:** While operating costs for electric aircraft are projected to be lower (due to cheaper "fuel" and less maintenance), the initial acquisition costs are high. proving the business case to operators and investors at scale remains a hurdle for many startups.

1.2 HISTORY OF ELECTRIFICATION IN AIRCRAFT

The history of electrification in aircraft is a story of continuous innovation driven by necessity, ambition, and environmental responsibility. From the rudimentary ignition circuits of early aircraft to today's megawatt-class electrical architectures, each era has pushed the boundaries of what is possible. Electrification has not only enhanced safety, efficiency, and performance but also redefined the future trajectory of aviation toward sustainability. As technology continues to evolve, the skies of tomorrow will likely be filled with aircraft powered not by combustion but by clean, silent, and efficient electric systems. The electrification of flight, once a futuristic vision, is now a central pillar in shaping the next century of aviation progress.

1.2.1 Early Beginnings: Mechanical and Pneumatic Dominance

In the early decades of flight, aircraft relied primarily on mechanical linkages, cables, and levers to control surfaces and systems. The Wright brothers' Flyer in 1903 had no electrical systems; all controls were mechanical, and propulsion was driven purely by a gasoline engine. As aviation advanced, particularly during World War I and II, the introduction of electrical systems was minimal and mainly limited to ignition systems, basic lighting, and simple communication devices.

Pneumatic and hydraulic systems dominated, especially for actuating landing gear, brakes, and flaps. By the 1930s, the use of small generators began to appear, providing limited electrical power to support radios and cockpit lighting. However, the overall dependency on non-electrical systems remained high due to limitations in electrical component reliability, weight, and efficiency. The concept of "electrification" in aircraft was still in its infancy, constrained by both technological immaturity and skepticism toward electric systems' performance under extreme aviation conditions.

1.2.2 World War II and the Birth of Electrical Integration

The Second World War became a critical turning point for aviation electrification. The rapid development of military aircraft demanded advanced control systems, reliable communication, and navigation aids. This period saw the introduction of electrical power systems capable of driving more complex instrumentation, radar, and weapon control systems. Aircraft such as the Boeing B-17 Flying Fortress and the Avro Lancaster featured alternators and generators to power extensive electrical networks, marking a significant departure from earlier designs. The wartime emphasis on performance and survivability accelerated the refinement of electrical components like generators, voltage regulators, and wiring insulation materials. These innovations set the stage for post-war civil aviation to incorporate electrification more extensively.

1.2.3 Jet Age and Electrical Power Expansion

The dawn of the jet age in the 1950s ushered in a new era for aircraft electrical systems. The introduction of high-speed, high-altitude flight placed new demands on power generation and distribution. Jet engines facilitated the installation of more robust electrical generators, capable of producing alternating current (AC) power rather than direct current (DC), which became the new standard for larger aircraft.

Commercial aircraft like the Boeing 707 and the Douglas DC-8 introduced more reliable and redundant electrical architectures, supporting advanced avionics, autopilot systems, and pressurization controls. The reliance on electricity began to expand beyond communication and lighting to critical flight systems. This period also marked the emergence of 115/200V AC, 400 Hz electrical standards, which remain foundational in modern aviation design. The increased use of electrical systems reduced dependence on hydraulics and pneumatics, gradually shifting aircraft architecture toward what would eventually become the "more-electric" aircraft concept.

Jet Revolution and the Spread of Electric Power

1.2.4 Rise of the "More-Electric Aircraft" Concept

Table: Evolution of Aircraft Electrification

Era	Major Development	Representative Aircraft
Early 1900s	Mechanical controls, minimal electrical systems	Wright Flyer (1903)
1930s–1940s	Introduction of generators for radios and lighting	Douglas DC-3, B-17
1950s–1960s	AC power systems, avionics integration	Boeing 707, DC-8
1970s–1990s	"More-Electric" concept, fly-by-wire controls	Airbus A320, Boeing 777
2000s–Present	Electric/hybrid propulsion, advanced energy storage	Airbus A350, Boeing 787, Eviation Alice

By the 1980s and 1990s, engineers began envisioning aircraft that would rely predominantly on electrical power to operate systems traditionally driven by mechanical, hydraulic, or pneumatic energy. The idea of the More-Electric Aircraft (MEA) was born, emphasizing efficiency, maintainability, and weight reduction. Electrical actuation replaced or augmented hydraulic systems for flight controls, brakes, and environmental management. For example, the Boeing 777, introduced in 1995, featured a highly integrated electrical system architecture with advanced fly-by-wire technology, electrical braking, and electrically driven hydraulic pumps. These innovations represented a crucial step toward the eventual goal of a fully electric or hybrid-electric aircraft.

1.2.5 Digital Age: Fly-by-Wire and Avionics Revolution

The introduction of fly-by-wire systems fundamentally transformed how aircraft are controlled. Instead of using mechanical linkages, these systems translate pilot inputs into electronic signals that control actuators and flight surfaces. The Airbus A320, launched in 1988, was the first commercial airliner to fully employ digital fly-by-wire technology, enabling superior precision, safety, and reduced pilot workload. Fly-by-wire represented the culmination of decades of electrical innovation, combining computer technology with reliable electrical systems to create the backbone of modern flight control. This development also set the stage for further integration of electrical components in propulsion, environmental control, and safety monitoring systems. Simultaneously, advances in avionics—ranging from flight management computers to GPS navigation and weather radar—depended heavily on electrical reliability. Electrification thus became inseparable from the digital transformation of aviation.

1.2.6 Emergence of Hybrid-Electric and All-Electric Aircraft

In the 21st century, environmental concerns and the push for sustainability have driven a new chapter in aircraft electrification. Manufacturers and research institutions are actively developing hybrid-electric and all-electric propulsion systems to reduce emissions, noise, and fuel consumption.

Aircraft such as the Boeing 787 Dreamliner embody the "More-Electric Aircraft" philosophy, replacing traditional pneumatic systems with electrically powered ones for de-icing, cabin pressurization, and engine start-up. The Dreamliner's electrical load exceeds 1 megawatt, nearly twice that of previous aircraft generations. Meanwhile, fully electric prototypes such as Eviation Alice, Pipistrel Alpha Electro, and Rolls-Royce Spirit of Innovation have demonstrated the potential for battery-powered flight. While range and energy density remain technical challenges, these milestones underscore a significant shift toward sustainable electrification.

1.2.7 Technological Milestones in Aircraft Electrification

The journey toward full electrification has been marked by several landmark innovations that have shaped the direction of aircraft design and operation.

- Introduction of electrical lighting and ignition systems (1910s).
- Development of onboard generators during World War II.
- Transition to 115/200V AC, 400 Hz systems in the 1950s.
- Integration of electric environmental control systems.
- Invention of solid-state power controllers in the 1970s.
- Implementation of digital fly-by-wire controls (1980s).

- Adoption of electrical braking and actuation (1990s).
- Emergence of the "More-Electric Aircraft" concept (2000s).
- Launch of the Boeing 787 with advanced electric systems (2011).
- Development of hybrid and fully electric aircraft prototypes (2020s).

Each of these advancements addressed specific engineering challenges, such as power distribution, weight optimization, and redundancy. Collectively, they reflect the broader trend of replacing mechanical complexity with electrical precision.

1.2.8 Challenges and Innovations in Electrification

Despite its advantages, the electrification of aircraft has presented significant engineering challenges. Chief among them are power density, thermal management, and reliability. Unlike ground-based systems, aviation demands lightweight, compact, and redundant electrical components capable of performing under extreme conditions. Battery technology, while rapidly improving, remains a limiting factor for fully electric flight due to its lower energy density compared to jet fuel. Researchers are exploring alternatives such as solid-state batteries, hydrogen fuel cells, and hybrid architectures that combine electrical propulsion with conventional engines. Additionally, advancements in power electronics, high-voltage DC distribution, and lightweight composite materials are helping overcome these limitations. Electric propulsion systems, if perfected, promise not only efficiency but also quiet operation, reduced emissions, and simplified maintenance—redefining the economics of aviation.

1.2.9 Modern Electrical Architectures and Smart Systems

Modern aircraft employ sophisticated electrical architectures that emphasize redundancy, smart distribution, and monitoring. For instance, the Boeing 787 and Airbus A350 feature distributed power electronics that allow selective isolation of faults and continuous power flow, enhancing safety. Digital monitoring systems also track the health of electrical components, predicting maintenance needs before failures occur—a concept known as predictive maintenance. These architectures support the seamless integration of renewable energy technologies and hybrid propulsion systems, paving the way for the next generation of green aviation.

1.3 DRIVERS OF ELECTRIFICATION IN AEROSPACE

1.3.1 Environmental

Electric aviation represents a beacon of hope for decarbonizing one of the most challenging sectors of the modern transportation system. While it is not a panacea, its potential to drastically reduce the climate impact and noise footprint of air travel is both significant and compelling. The market is poised for explosive growth, driven by regulatory pressures, technological advances, and a growing collective commitment to environmental stewardship.

However, the journey ahead is complex, requiring breakthroughs in battery technology, the development of extensive new infrastructure, and the creation of supportive international policies and regulations. The pioneering electric aircraft already in the skies today are not merely prototypes; they are the vanguard of a profound industrial shift. By continuing to invest in research, foster collaboration, and implement smart policies, the aviation industry can navigate the challenges ahead and fully unlock the potential of electric flight. In doing so, it will secure a cleaner, quieter, and more sustainable future for global connectivity, ensuring that the benefits of air travel can be enjoyed by generations to come without compromising the health of our planet.

Environmental Benefits of Electric Aviation

The primary environmental advantage of electric aircraft lies in their capacity to eliminate in-flight emissions. Unlike conventional aircraft that burn jet fuel, releasing carbon dioxide, nitrogen oxides, and other pollutants directly into the atmosphere, fully electric aircraft powered by batteries produce zero emissions at the point of use. This direct elimination of combustion processes addresses the core of aviation's air quality and climate impact. Research from academic institutions underscores this benefit, revealing that a small electric aircraft can have a up to 60% lower climate impact over its life cycle compared to its fossil-fueled equivalent, provided it is charged using green electricity. This substantial reduction highlights the significant potential for electric aviation to mitigate the sector's contribution to global warming.

Beyond greenhouse gas emissions, electric aviation offers a dramatic reduction in noise pollution. The electric motors used in modern aircraft operate at significantly lower noise levels compared to traditional turboprop or jet engines. This reduction opens the possibility of operations with less community disturbance, potentially allowing for flights earlier in the morning or later in the evening without noise-related complaints.

Furthermore, Life Cycle Assessment studies provide a comprehensive view of the environmental trade-offs. While electric aircraft excel in reducing airborne emissions, they also shift a portion of the environmental burden to other phases, notably the manufacturing and disposal of batteries, which involves resource-intensive processes and the use of scarce minerals. This holistic perspective is crucial for ensuring that the transition to electric aviation delivers genuine, system-wide environmental benefits rather than simply displacing impacts.

Market Growth

The electric aircraft market is experiencing exponential growth, signaling strong confidence from industry stakeholders and investors in its future. The global market size is predicted to expand significantly over the coming decade, reflecting a robust compound annual growth rate. This remarkable growth trajectory is fueled by a confluence of powerful drivers that are accelerating the adoption of electric aviation technologies across the globe.

- **Stringent Environmental Regulations**: Global commitments such as the aviation industry's goal to achieve net-zero carbon emissions by 2050 are creating a powerful regulatory push for cleaner technologies.
- **Economic Factors**: Volatile and rising fuel prices are making the superior energy efficiency of electric propulsion increasingly attractive from an operational cost perspective.
- **Technological Breakthroughs**: Advances in battery energy density, lightweight composite materials, and electric propulsion systems are enhancing the performance and feasibility of electric aircraft.
- **Consumer Demand**: A shift in consumer sentiment towards more sustainable travel options is generating demand for greener alternatives, prompting airlines to invest in electric fleets.
- **Regional Initiatives**: North America leads the market due to robust aviation infrastructure and significant R&D investments, while Europe is projected to be the fastest-growing region, driven by environmental initiatives and urban air mobility projects.
- **Corporate Investment**: Major aerospace companies and startups are allocating substantial resources to electric aircraft development, signaling long-term commitment to the sector.
- **Infrastructure Development**: Governments and private entities are beginning to invest in the necessary charging and support infrastructure for electric aircraft.
- **Urban Air Mobility**: The growing urban air mobility market is creating additional demand for electric vertical takeoff and landing aircraft.
- **Regional Connectivity**: Electric aircraft offer economic solutions for enhancing regional connectivity while maintaining environmental standards.
- **Competitive Advantage**: Airlines and operators see early adoption of electric aircraft as a potential source of competitive advantage in an increasingly environmentally-conscious market.

Widespread Adoption

Despite the optimistic forecasts and compelling benefits, the path to widespread electric aviation is fraught with significant challenges that must be overcome. A primary obstacle is the current state of battery technology. The energy density of today's lithium-ion batteries remains insufficient for longer-haul flights. This limitation directly restricts the range and payload capacity of electric aircraft, confining them largely to short-haul, regional routes. While batteries are improving, they also have a finite lifespan and present environmental concerns related to their production and disposal, creating a trade-off between reducing climate impact and increasing mineral resource scarcity.

The adoption of electric aviation is also hampered by a lack of supporting infrastructure and regulation. Airports worldwide will require extensive and costly upgrades to support electric aircraft, including the installation of high-capacity charging stations and upgrades to local energy grids to handle the substantial power demand. Furthermore, regulatory frameworks, both national and international, are struggling to keep pace with technological innovation. Current air service regulations are often too rigid to accommodate new aircraft types, and there is a pressing need for common international standards for airport infrastructure and aircraft certification to ensure safety and interoperability. Politically, there is a need for more unified strategies and committed funding. Various regions have identified a lack of political commitment and uncoordinated approaches as barriers to implementing electric aviation. Finally, the economic viability of early-generation electric aircraft remains uncertain, with high initial costs and potentially high-ticket prices posing a risk to market growth until economies of scale are achieved.

Leading Electric Aircraft and their Environmental Role

The electric aviation revolution is being pioneered by a diverse fleet of innovative aircraft, each demonstrating unique approaches to sustainable flight. These pioneering models serve as critical testbeds for technology and provide tangible proof of concept for various market segments. The Pipistrel Velis Electro stands as the first type-certified electric aircraft, serving primarily as a training aircraft that demonstrates the practical application of electric propulsion while significantly reducing noise and emissions during flight training operations. This aircraft has proven that electric flight is not merely a theoretical concept but a practical reality for certain aviation segments.

The Eviation Alice represents a step forward in commercial applications, designed as a nine-passenger commuter aircraft that utilizes advanced composite materials to maximize efficiency. Its all-electric design targets regional travel markets, offering a glimpse into the future of short-haul air travel without direct emissions. Similarly, the Heart Aerospace ES-30 takes a hybrid-electric approach, accommodating up to 30 passengers and addressing the range limitations of purely electric aircraft while still offering significant emission reductions on shorter routes. This aircraft has garnered substantial interest from major airlines, indicating industry confidence in the hybrid-electric pathway.

Other notable aircraft include the Beta Technologies ALIA CX300, which emphasizes cargo operations and demonstrates the cross-country capability of electric aircraft through extensive flight testing. These various aircraft designs collectively contribute to the advancement of electric aviation by addressing different market needs while sharing the common goal of reducing aviation's environmental impact. Each model provides valuable data and operational experience that informs the development of future generations of electric aircraft, creating a foundation for continuous improvement in performance, reliability, and sustainability.

Road Ahead for Electric Aviation

The future trajectory of electric aviation hinges on continued and synchronized progress across multiple domains. Technological innovation remains the cornerstone, with intensive research and development needed to overcome current limitations. Next-generation battery chemistries hold the promise of dramatically higher energy densities and longer lifespans, which would directly translate into increased range and payload for electric aircraft. Furthermore, the integration of Artificial Intelligence is set to play a pivotal role in optimizing energy management, predicting battery degradation, and enhancing aerodynamics through generative design, thereby improving overall efficiency, safety, and performance.

Equally important is the establishment of a supportive policy and infrastructure ecosystem. The creation of a unified international strategy and harmonized regulatory standards is crucial for fostering cross-border electric aviation and providing the certainty needed for long-term investment.

Governments and industry bodies must collaborate to fund and modernize airport infrastructure, ensuring it can support the unique demands of electric aircraft charging and operations. The journey toward widespread adoption will likely be phased. The industry will initially focus on short-haul regional routes, air taxis, and cargo operations, which are less constrained by current battery limitations. As technology matures, these applications will serve as stepping stones toward the ultimate goal of larger, and eventually long-haul, electric commercial flight.

- **Battery Technology Advancement**: Development of higher energy density batteries through new chemistries and manufacturing processes.

- **Charging Infrastructure**: Establishment of rapid charging systems at airports compatible with various aircraft types.
- **Grid Modernization**: Upgrades to electrical grids to support high-power charging demands at airport facilities.
- **Regulatory Harmonization**: Development of international standards for certification, operation, and maintenance of electric aircraft.
- **Workforce Training**: Creation of educational programs and certification standards for technicians and pilots specializing in electric aircraft.
- **Sustainable Energy Integration**: Coupling airport charging infrastructure with renewable energy sources to maximize environmental benefits.
- **Noise Abatement Procedures**: Development of flight procedures that leverage the quiet operation of electric aircraft.
- **Lifecycle Management**: Establishment of recycling and repurposing systems for aircraft batteries.
- **Air Traffic Management Integration**: Adaptation of air traffic control systems to accommodate the unique characteristics of electric aircraft operations.
- **Public Acceptance**: Education and demonstration programs to build trust and familiarity with electric aircraft technology among the traveling public.

1.3.2 Economic

Electric aviation's economics are not a simple fuel-swap story; they represent a re architecture of the cost base, revenue logic, and supporting infrastructure of flight. The strongest early cases will combine short sectors, high frequency, premium time value, and energy-savvy operations—conditions found in urban air mobility corridors, regional island hops, and select cargo lanes. As certification milestones are met and fleets scale, learning curves in batteries and power electronics should bend costs downward, while grid investments and standardized charging make operations more predictable.

Policy can accelerate the crossover point by properly valuing lower emissions and noise. The next decade will not eliminate conventional aviation, but it will redraw the short-haul map: more routes, quieter communities, and business models built around electrons, data, and high-utilization assets. Those who treat batteries and infrastructure as core economic levers—not afterthoughts—will capture the earliest and most durable value.

Cost Structure and Unit Economics

The economic promise of electric flight rests on a different cost stack: more upfront capital investment in batteries and power systems, offset by materially lower operating expenses. Electric motors have high energy conversion efficiency and far fewer moving parts than internal combustion engines or gas turbines, reducing maintenance hours and spare-part inventories. Energy costs also tilt in electricity's favor on a per-unit-of-energy basis, especially when charged off-peak or contracted via long-term power purchase agreements.

However, energy is only part of the story. Batteries add weight that reduces revenue payload, degrade over cycles, and impose replacement costs that behave like an additional consumable. The result is a delicate balance: short stage lengths, high utilization, and optimized charging routines can drive attractive cost per available seat-mile, while longer routes or underutilized aircraft can erode the advantage. Early operators will focus obsessively on turn times, charging logistics, and dispatch reliability to protect margins.

Battery Economics and Lifecycle

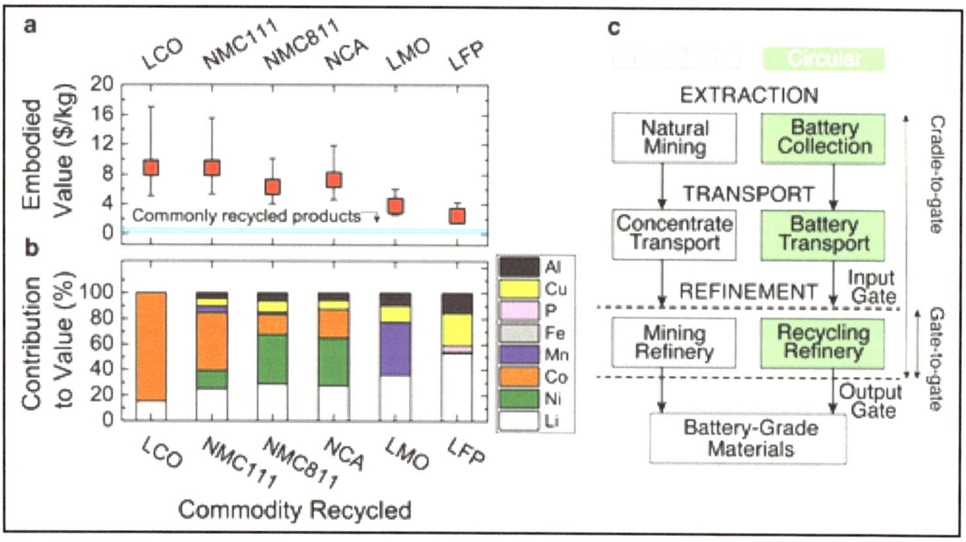

Fig: Battery Cost Efficiency and Operational Life

Battery packs are both asset and consumable, and the way they are financed, operated, and replaced will define electric aviation's cost curve. Pack-level energy density dictates range and payload; cycle life and calendar aging dictate amortized cost per flight hour. Even as automotive pack costs approach mass-market levels, aviation packs must meet stricter certification, safety, and thermal requirements, often at higher $/kWh. The economic lever is optimizing depth of discharge, charge rates, and thermal conditions to extend cycle life without sacrificing turnaround time. A battery's residual value—whether as a second-life storage asset or via materials recycling—also matters, lowering effective depreciation. Operators will increasingly treat batteries as modular, swappable assets with distinct financing from the airframe, enabling "battery-as-a-service" arrangements that smooth cash flows and transfer technology obsolescence risk to specialized providers.

Infrastructure, Energy Pricing, and Turnaround

Electric aircraft economics hinge on what happens at the gate as much as in the air. Airports and vertiports need dependable megawatt-scale charging, grid interconnections sized to peak loads, and, often, on-site storage to arbitrage time-of-use rates and buffer against grid constraints. The business case improves when charging can be scheduled during off-peak hours, paired with behind-the-meter solar, or contracted under fixed-price tariffs.

Turnaround discipline is crucial: every minute plugged in rather than flying reduces asset utilization. Fast charging shortens ground time but accelerates battery wear; slower charging preserves batteries but risks schedule delays. The financially optimal strategy will vary by route network, load factors, and local electricity pricing. Airports, in turn, will explore new revenue streams—charging services, power resale, and demand response—becoming energy hubs as well as transport nodes.

Market Segments and Use Cases

Not all flying is created equal economically. Near-term viability concentrates where battery constraints align with real demand. On the urban side, electric vertical takeoff and landing (eVTOL) aircraft target short intra-city or regional hops that substitute for congested car trips, selling time savings and network reliability. Their unit economics depend on high-frequency operations along well-chosen corridors, premium pricing, and tight turnaround routines. On the regional side, fixed-wing electric or hybrid-electric aircraft serve 100–300 km routes, revitalizing thin air corridors where conventional jets are uneconomic. These aircraft can unlock previously marginal markets by slashing operating costs and noise footprints, enabling service to smaller airports with lower fees and fewer curfews. Cargo and medical logistics add further niches with high willingness to pay, predictable schedules, and less stringent comfort requirements.

Financing, Business Models, and Ecosystem

Electric aviation encourages unbundling. Airframe OEMs may sell aircraft at lower margins while monetizing software, performance upgrades, and energy management. Lessors could finance airframes and batteries separately, reflecting different depreciation curves and residual value risks. Operators may sign long-term electricity contracts, hedge power costs with on-site generation, or outsource charging to airport-owned energy companies. Fleet management will rely on predictive maintenance driven by motor telemetry and battery health analytics, shrinking unplanned downtime. Ecosystem partnerships—OEMs with utilities, airports with charging providers, cities with vertiport developers—help align capital deployment and reduce stranded assets. Ultimately, the winning models will de-risk three things: technology obsolescence, regulatory timelines, and demand ramp, while maintaining high aircraft utilization and predictable cash yields.

Policy, Externalities, and Community Benefits

Policy can tip marginal routes into profitability and speed infrastructure rollouts. Carbon pricing, clean energy credits, and accelerated depreciation for low-emission assets improve total cost of ownership for electric aircraft. Noise regulations and community acceptance increasingly constrain traditional aviation; quieter electric operations can secure operating windows, unlock new routes, and reduce land-use conflicts—economic benefits not always captured in simple cost-per-seat metrics.

Public procurement and regional air service subsidies can catalyze early deployments in thin markets, seeding demand where commercial cases are close to break-even. Transparent lifecycle accounting—including upstream battery materials and electricity carbon intensity—will determine eligibility for incentives and shape public trust. As grids decarbonize, the climate advantage of electric aviation strengthens, further widening the policy-supported cost gap with fossil-fueled alternatives.

Global and Regional Perspectives

Economics vary widely by region. Markets with high fuel prices, carbon taxes, and dense short-haul demand—think Europe and parts of Asia—offer earlier break-even opportunities. Regions with constrained grids or highly seasonal traffic may lag unless paired with storage and renewables. In emerging markets, short-runway or water-landing electric seaplanes could improve regional connectivity at lower operating costs, but financing and maintenance ecosystems remain hurdles. Supply chains matter: access to battery materials, certified maintenance centers, and skilled technicians influences both cost and uptime. Cross-border harmonization of certification standards will lower financing costs, expand residual markets for used aircraft and batteries, and accelerate learning curves across the industry.

Risks, Timelines, and Learning Curves

Two clocks are ticking: certification and cost decline. Certification is capital intensive and timelines can slip, delaying revenue and straining balance sheets. Cost declines follow learning curves—batteries, motors, and inverters typically get cheaper and better with cumulative production—but aviation volumes start low, tempering early gains. Meanwhile, competing decarbonization pathways like sustainable aviation fuel and hydrogen may capture policy support or customer budgets on overlapping routes. Key risks include slower-than-expected battery improvements, grid bottlenecks that raise energy prices at airports, and utilization shortfalls from regulatory or community constraints. Mitigations include hybrid-electric architectures for longer routes, modular battery bays, diversified revenue lines (cargo, medical, tourism), and flexible charging strategies. Over time, compounding improvements in energy density, cycle life, software, and operations can push electric aviation from niche to mainstream on short-haul missions.

Ten Economic Levers to Watch

- Battery pack cost, energy density, and cycle life, which determine range, payload, and amortized cost per flight.
- Electricity procurement strategy, including time-of-use arbitrage, PPAs, and on-site renewables plus storage.
- Charging speed and turnaround optimization to maximize aircraft utilization without accelerating battery wear.
- Maintenance savings from simpler drivetrains versus new inspection requirements for high-voltage systems.
- Route selection that matches battery-constrained range to high-demand, time-sensitive corridors.
- Financing structures separating airframe and battery assets to align depreciation and residual value risks.
- Policy incentives, carbon pricing, and noise-based operating windows that change revenue potential.
- Airport and vertiport infrastructure readiness, grid capacity, and the economics of third-party charging.
- Software-enabled operations: demand forecasting, dispatch, and predictive maintenance that lift load factors.
- Competitive landscape with SAF and hybrid solutions, shaping price benchmarks and adoption pacing.

1.3.3 Technological

The technological scope of electric aviation is vast, moving far beyond simply swapping a fuel tank for a battery. It encompasses a complete reimagining of how aircraft are powered, controlled, and constructed. While current technologies have successfully enabled small, short-range trainers and nascent urban air mobility vehicles, scaling these to larger commercial aircraft requires breakthrough advancements in energy storage chemistry, high-voltage power distribution, and thermal management. The industry is currently in a phase of rapid, iterative development, where established aerospace giants and agile startups alike are racing to mature these varied technological pillars. As these technologies converge and mature, they promise to deliver an era of aviation that is not only cleaner and quieter but fundamentally more efficient in its utilization of energy.

Challenge of Energy Storage

The single most critical technological hurdle in electric aviation is energy storage. Traditional jet fuel is extraordinarily energy-dense, containing roughly 12,000 watt-hours per kilogram (Wh/kg). In stark contrast, current state-of-the-art certified lithium-ion battery packs for aviation generally offer between 150 and 250 Wh/kg at the pack level. This massive disparity means that for a given range, an electric aircraft must carry a significantly higher proportion of its maximum takeoff weight as "fuel" (batteries), severely limiting payload and endurance.

Consequently, immense research is focused on increasing this specific energy. While lithium-ion chemistries are the current standard due to their proven reliability, they are approaching their theoretical performance limits. The industry's technological roadmap heavily relies on the eventual commercialization of next-generation chemistries, such as lithium-sulfur or solid-state batteries, which promise to double or triple current energy densities while potentially offering improved safety profiles by eliminating flammable liquid electrolytes.

Electric Propulsion Units (EPUs)

While energy storage remains a bottleneck, the technology for converting that energy into thrust—the Electric Propulsion Unit (EPU)—is highly advanced and offers significant advantages over internal combustion engines. Electric motors are incredibly efficient, typically converting over 90% of input electrical energy into mechanical work, compared to roughly 30-40% for piston engines or small turbines. Furthermore, electric motors provide instantaneous torque, are mechanically simpler with far fewer moving parts, and do not suffer from power loss at higher altitudes to the same degree as naturally aspirated combustion engines. The technological focus in this area is on improving "power density"—the amount of power a motor can generate relative to its weight and size. Innovations in motor topology, such as axial flux motors which are thinner and lighter than traditional radial flux designs, are critical for aerospace applications where every kilogram counts.

Core Technological Pillars and Innovations

The realization of practical electric aircraft requires varied, interdependent technologies working in perfect unison. To understand the depth of the engineering required, one must look at the specific technological components that differentiate these vehicles from their predecessors. The following ten technological pillars are currently defining the development of electric aviation:

- **Lithium-Ion Battery Architectures:** Currently the baseline technology for all-electric aircraft, these systems require sophisticated Battery Management Systems (BMS) to monitor individual cell voltages, temperatures, and state of health to prevent thermal runaway events in flight.
- **Solid-State Batteries:** Considered the "holy grail" for future electric flight, this technology replaces the liquid electrolyte with a solid material, potentially doubling energy density and significantly improving fire safety, though they currently face challenges in manufacturing scale and cycle life.
- **Axial Flux Motors:** These motors align magnetic flux parallel to the axis of rotation (rather than radially), allowing for compact, high-torque, and lightweight designs that are ideally suited for driving propellers directly without heavy gearboxes.
- **Silicon Carbide (SiC) Inverters:** Essential for converting DC power from batteries to AC power for motors, SiC-based power electronics can operate at higher voltages and temperatures with greater efficiency than traditional silicon-based components, reducing cooling needs.
- **Distributed Electric Propulsion (DEP):** This aerodynamic technology involves using multiple small electric motors spread along the wing leading edge, increasing lift at low speeds and allowing wings to be smaller and more efficient for cruise flight.
- **Thermal Management Systems:** High-voltage electric powertrains generate intense localized heat during takeoff and climb; advanced liquid cooling or phase-change material systems are technologically vital to dissipate this heat without adding excessive drag or weight.
- **High-Voltage Architectures:** To transport the massive amounts of power required for flight without using prohibitively heavy, thick copper cables, aircraft are moving to much higher voltage standards (often 800V to 1000V+), requiring new standards for insulation and arc-fault protection.
- **Hydrogen Fuel Cell Systems:** A parallel technological path that converts hydrogen gas directly into electricity and water; this technology offers much higher energy density than batteries for longer-range flights but requires complex cryogenic storage tanks and thermal management for the fuel cells.
- **Turbogenerator Hybrids:** For larger aircraft, existing battery tech is insufficient. Series-hybrid technologies use a highly efficient gas

turbine optimized to run at a steady state solely to generate electricity, which then powers electric motors or charges batteries, bridging the gap between pure combustion and all-electric.

- **Advanced Lightweight Composites:** Because batteries are so heavy, the rest of the airframe must be lighter than ever. The use of advanced carbon fiber reinforced polymers (CFRP) and novel manufacturing techniques is essential to offset propulsion system weight and maintain a viable payload.

Power Electronics and Systems Integration

The "nervous system" of an electric aircraft is its power electronics. Managing megawatt-scale power flows in a vibrating, altitude-variable environment is a profound technological challenge. The inverter, which converts Direct Current (DC) from batteries into Alternating Current (AC) for the motors, must be exceptionally reliable; a failure here is equivalent to an engine failure in a traditional aircraft. Furthermore, as voltages increase to reduce cable weight, the risk of electrical arcing at high altitudes (where thinner air provides less insulation) increases.

This phenomenon, known as Paschen's Law, requires engineers to develop new forms of insulated wiring and shielded connectors specifically for high-voltage aerospace applications. Integrating these complex systems so they communicate flawlessly—ensuring that a pilot's throttle command translates instantly and accurately to motor RPM while simultaneously managing battery draw and cooling pumps—requires highly sophisticated software and redundant control architectures.

Hydrogen-Electric Alternatives

Recognizing the physical limitations of battery technology for larger, longer-range aircraft, a significant portion of technological development is focused on hydrogen-electric powertrains. In these systems, hydrogen (stored either as a highly compressed gas or a cryogenic liquid at -253°C) is fed into a fuel cell stack. Through an electrochemical reaction with oxygen from the air, electricity is produced to drive the same electric motors used in battery aircraft.

The technological challenges here shift from energy density to volume and storage. Liquid hydrogen is light but requires large, heavily insulated tanks that can compromise aircraft interior space or aerodynamics. Moreover, the fuel cells themselves can be heavy and require significant cooling. Despite these hurdles, hydrogen technology currently offers the only viable all-electric pathway for regional or single-aisle class airliners in the coming decades.

1.4 ELECTRIC VS. CONVENTIONAL PROPULSION: KEY DIFFERENCES

The comparison between electric and conventional propulsion highlights a technological crossroads in aviation's evolution. Conventional propulsion remains dominant due to its proven reliability, range, and infrastructure, but it faces mounting pressure from environmental regulations and rising fuel costs. Electric propulsion, with its promise of cleaner, quieter, and more efficient flight, represents the future of sustainable aviation.

While current limitations in battery technology and energy density restrict electric aircraft to shorter missions, rapid progress in research suggests that these barriers will diminish over the coming decades. Ultimately, the coexistence of both propulsion types—conventional for long-haul and electric for regional and urban transport—will shape the future landscape of aviation. The transition toward electric propulsion is not just a technological advancement but a vital step toward achieving a more sustainable and environmentally conscious era of flight.

1.4.1 Fundamentals of Conventional Propulsion

Conventional propulsion in aircraft is based on the combustion of fossil fuels such as aviation gasoline (avgas) or jet fuel to generate mechanical power. The core components of traditional propulsion systems include the combustion chamber, turbines, compressors, and exhaust systems. These systems convert chemical energy stored in the fuel into kinetic energy, producing thrust that propels the aircraft forward. The most common types of conventional propulsion are piston engines and gas turbines.

Piston engines, used in smaller aircraft, operate on the same principle as automobile engines, with reciprocating motion driving propellers. In contrast, gas turbines and turbojets, used in larger and faster aircraft, use continuous combustion and turbine-driven compressors to create high-speed exhaust gases that generate thrust. Although conventional propulsion has proven reliability, it is energy-intensive and environmentally costly, producing high levels of carbon dioxide, nitrogen oxides, and particulates. The efficiency of these systems depends on engine design, altitude, and speed, but the reliance on fossil fuels makes them inherently unsustainable in the long term.

1.4.2 Principles of Electric Propulsion

Electric propulsion represents a significant technological departure from traditional engines. Instead of burning fuel, it relies on electrical energy—typically stored in batteries, ultracapacitors, or hydrogen fuel cells—to drive electric motors that generate thrust. The core mechanism of electric propulsion involves the conversion of electrical energy into mechanical motion through electromagnetic forces within the motor. Electric propulsion systems can be broadly classified into battery-electric, hybrid-electric, and fuel cell-electric configurations. In a battery-electric setup, energy stored in lithium-ion or solid-state batteries powers the motors directly. Hybrid-electric systems combine conventional engines with electric drives, allowing improved efficiency and reduced emissions. Fuel cell systems, meanwhile, use hydrogen and oxygen to produce electricity via electrochemical reactions, offering longer range potential without combustion. The absence of combustion and mechanical complexity results in quieter, cleaner, and potentially more efficient operations. However, the primary limitation of electric propulsion lies in energy storage, as current batteries have significantly lower energy density compared to liquid fuels.

1.4.3 Energy Density and Efficiency Considerations

One of the most critical differences between electric and conventional propulsion lies in energy density, or the amount of energy stored per unit weight. Jet fuel, with an energy density of about 43 MJ/kg, vastly outperforms even the most advanced lithium-ion batteries, which store around 1 MJ/kg. This gap means that electric aircraft currently have shorter ranges and smaller payload capacities compared to their conventionally powered counterparts.

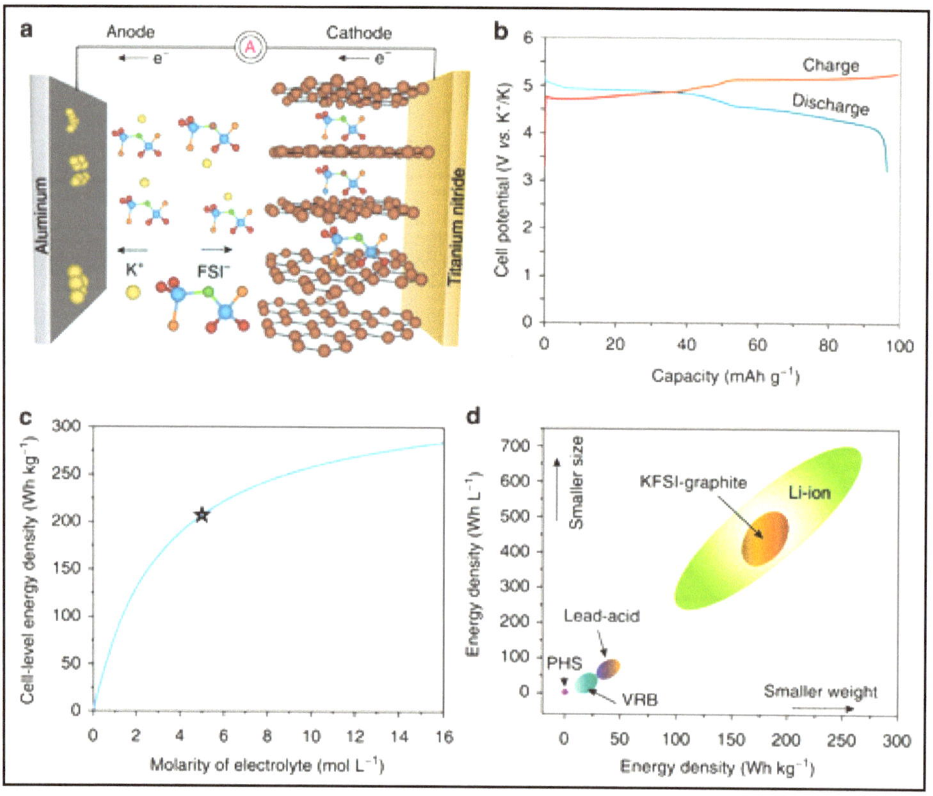

Fig: Energy Concentration and Performance Efficiency Factors

Despite this, electric propulsion systems are inherently more efficient in energy conversion. Conventional combustion engines typically achieve 30–40% thermal efficiency, losing most of their energy as heat. In contrast, electric motors can reach efficiencies exceeding 90%, meaning they convert almost all input energy into useful work. The challenge, therefore, lies not in the efficiency of the motor itself but in the energy storage and power delivery systems that support it.

Table: Comparison between Electric and Conventional Propulsion

Parameter	Electric Propulsion	Conventional Propulsion
Energy Source	Batteries, fuel cells, hybrid systems	Aviation gasoline or jet fuel
Efficiency	85–95%	30–40%
Emissions	Zero or very low	High CO_2 and NO_x emissions
Noise	Low operational noise	High engine and exhaust noise
Maintenance	Minimal (few moving parts)	Complex, requires frequent servicing
Range	Limited by battery capacity	Long, depending on fuel storage
Weight Impact	Batteries increase weight	Fuel decreases as consumed
Cost per Flight Hour	Lower energy cost	Higher fuel cost
Reliability	Dependent on battery performance	Established and proven technology
Sustainability	High potential	Environmentally unsustainable

1.4.4 Performance and Operational Differences

The operational performance of electric and conventional propulsion systems diverges considerably in areas such as range, power output, and altitude performance. Conventional propulsion excels in long-distance travel and high-speed operations, making it suitable for commercial and military aviation. Jet engines are optimized for efficiency at high altitudes, where thinner air reduces drag and enhances thrust efficiency. Electric propulsion, however, is better suited for short-haul flights, regional transport, and urban air mobility. Electric motors deliver instant torque, enabling faster acceleration and smoother operation. Moreover, electric propulsion allows distributed propulsion configurations—multiple smaller motors placed along the airframe—which can improve aerodynamic efficiency and redundancy. However, as battery energy density improves, electric aircraft will begin to close the gap in range and endurance. Emerging technologies, such as solid-state batteries and hydrogen fuel cells, aim to increase flight durations while maintaining the environmental benefits of electrification.

1.4.5 Environmental Impact and Sustainability

The most profound difference between electric and conventional propulsion lies in their environmental impact. Conventional engines emit greenhouse gases and pollutants that contribute to global warming and air quality degradation. The aviation industry is responsible for around 2–3% of global CO_2 emissions, a figure expected to rise without significant technological change. Electric propulsion, on the other hand, offers a path to near-zero-emission flight, particularly when powered by renewable energy sources. Battery-electric and hydrogen-powered aircraft produce no in-flight emissions, significantly reducing environmental footprints. Additionally, electric systems generate much less noise pollution, an increasingly important factor for urban and regional air travel. Yet, the sustainability of electric propulsion depends on the life cycle of the energy source. If the electricity used to charge batteries or produce hydrogen comes from fossil fuels, the overall environmental benefit diminishes. Therefore, the future success of electric propulsion is tied closely to the global shift toward renewable energy generation.

1.4.6 Economic and Maintenance Considerations

Economic factors also highlight notable contrasts between the two propulsion types. Conventional aircraft incur high operational costs due to fuel expenses, maintenance, and engine overhauls. The mechanical complexity of turbines and piston engines means frequent inspection and replacement of components, which drives up maintenance costs and downtime. Electric propulsion systems, in contrast, are mechanically simpler, with fewer moving parts and lower wear-and-tear rates. This simplicity translates into reduced maintenance requirements and longer component lifespans. Additionally, electricity—especially from renewable sources—is cheaper and more stable in price compared to aviation fuel, lowering the overall cost per flight hour. However, the initial acquisition costs of electric aircraft can be higher due to expensive battery systems and new infrastructure needs, such as high-capacity charging or hydrogen refueling stations. Over time, technological maturity and mass production are expected to drive costs down, making electric propulsion economically competitive.

1.4.7 Differences between Electric and Conventional Propulsion

- **Energy Source:** Electric systems use stored electrical energy, while conventional engines burn fossil fuels.
- **Efficiency:** Electric motors achieve higher conversion efficiency compared to combustion engines.
- **Range:** Conventional propulsion supports long-distance flights; electric systems are limited by battery life.
- **Maintenance:** Electric systems require less maintenance due to fewer mechanical components.

- **Environmental Impact:** Electric aircraft produce zero emissions, whereas conventional engines emit CO_2 and NO_x.
- **Noise Levels:** Electric propulsion is significantly quieter during operation.
- **Infrastructure Needs:** Conventional systems rely on established fuel networks; electric systems need new charging infrastructure.
- **Energy Density:** Jet fuel stores far more energy per kilogram than current batteries.
- **Operational Cost:** Electric aircraft offer lower running costs but higher initial investment.
- **Technological Maturity:** Conventional propulsion is fully developed and certified, while electric technology is still emerging.

These differences reveal that while electric propulsion is promising, its large-scale adoption depends on overcoming limitations in energy storage, infrastructure, and certification frameworks.

1.5 OVERVIEW OF INDUSTRY LANDSCAPE AND MARKET TRENDS

1.5.1 Macro Trends Reshaping Industries

Several powerful macro trends are simultaneously transforming industry landscapes across the globe. Digital transformation continues to be the dominant force, with organizations accelerating their adoption of digital technologies to enhance operational efficiency, improve customer experiences, and create new value propositions.

The COVID-19 pandemic served as a significant catalyst, compressing years of digital adoption into months and establishing remote work capabilities as a standard rather than an exception. Sustainability and environmental, social, and governance considerations have evolved from corporate social responsibility initiatives to core business imperatives. Consumers, investors, and regulators are increasingly demanding that companies demonstrate genuine commitment to sustainable practices, circular economy principles, and ethical governance.

Demographic shifts, including aging populations in developed economies and youth bulges in emerging markets, are creating divergent challenges and opportunities across regions. The changing nature of work, accelerated by automation and artificial intelligence, is forcing organizations to rethink talent strategies, workplace structures, and skill development programs. These macro trends are not operating in isolation but are interacting in complex ways to reshape competitive landscapes across virtually every industry sector.

- **Digital Transformation Acceleration:** Organizations are prioritizing cloud migration, data analytics, and digital workflow implementation to enhance resilience and operational flexibility.
- **Sustainability Integration:** Companies are embedding environmental considerations into core business strategies, driven by investor pressure, consumer demand, and regulatory requirements.
- **Supply Chain Reconfiguration:** Businesses are diversifying suppliers, increasing inventory buffers, and leveraging technology to build more resilient and transparent supply networks.
- **Remote and Hybrid Work Models:** The normalization of distributed workforces is transforming workplace policies, real estate strategies, and collaboration technologies.
- **Artificial Intelligence Proliferation:** AI and machine learning applications are moving from experimental projects to core operational components across functions.
- **Changing Consumer Behaviors:** Customers increasingly expect personalized, seamless, and ethical experiences across physical and digital touchpoints.
- **Geopolitical Realignment:** Companies are adapting to changing trade relationships, regulatory divergence, and regional economic alliances.
- **Talent Market Transformation:** Organizations are rethinking attraction, retention, and development strategies in response to shifting employee expectations and skill requirements.
- **Platform Business Model Expansion:** Platform-based approaches continue to disrupt traditional linear business models across multiple industries.
- **Health and Wellbeing Prioritization:** The pandemic has elevated health considerations into mainstream business planning and employee value propositions.

1.5.2 Technology Sector Landscape and Evolution

The technology industry continues to be the primary engine of disruption and innovation across the global economic landscape. Cloud computing has matured from a cost-saving measure to a strategic enabler of business agility and digital innovation, with hybrid and multi-cloud architectures becoming the standard for enterprise organizations. The artificial intelligence and machine learning market has entered a phase of rapid commercialization, with practical applications emerging in areas such as predictive analytics, natural language processing, and computer vision. The convergence of AI with other technologies like Internet of Things and blockchain is creating new possibilities for autonomous systems and trusted transactions. Cybersecurity has evolved from a technical concern to a core business risk, driving increased investment in threat detection, identity management, and zero-trust architectures. The semiconductor industry, once a background enabler, has moved to center stage due to supply chain constraints and geopolitical tensions, prompting massive investment in new fabrication capacity and regional self-sufficiency initiatives. The technology landscape is characterized by both consolidation among major platforms and vibrant innovation from startups, particularly in emerging areas such as quantum computing, edge computing, and Web3 technologies that promise to define the next wave of digital transformation.

1.5.3 Healthcare Industry Transformation

The healthcare industry is undergoing a fundamental restructuring driven by technological innovation, changing consumer expectations, and regulatory evolution. The pandemic accelerated the adoption of telehealth and remote patient monitoring, establishing digital health as a permanent component of care delivery models.

Pharmaceutical companies are leveraging artificial intelligence and advanced analytics to accelerate drug discovery and development, while simultaneously facing increased pressure on drug pricing and public scrutiny of their role in public health. Medical device manufacturers are incorporating connectivity and data analytics into their products, creating new opportunities for preventive care and personalized treatment. The healthcare landscape is shifting toward value-based care models that prioritize patient outcomes over service volume, forcing providers to develop new capabilities in data analysis, population health management, and care coordination.

Digital health startups are challenging traditional healthcare delivery models with convenient, consumer-focused solutions for everything from mental health to chronic disease management. Meanwhile, healthcare systems are grappling with workforce challenges, cost pressures, and the need to address health inequities that have been highlighted and exacerbated by the pandemic, creating both challenges and opportunities for innovation across the sector.

1.5.4 Retail and Consumer Goods Evolution

The retail and consumer goods sector has experienced more transformation in the past decade than in the previous century, with changing consumer behaviors and digital technologies rewriting the rules of competition. E-commerce continues to gain share, but the landscape has evolved beyond simple online transactions to encompass omnichannel experiences that seamlessly blend physical and digital touchpoints. The rise of direct-to-consumer brands has disrupted traditional distribution channels and forced established manufacturers to develop new capabilities in digital marketing, data analytics, and customer relationships. Sustainability has become a significant differentiator, with consumers increasingly considering environmental impact, ethical sourcing, and corporate values in their purchasing decisions.

Personalization has moved from a competitive advantage to a baseline expectation, driven by advances in data analytics and artificial intelligence. The physical store is being reimagined as an experiential destination and fulfillment hub rather than just a point of sale. Supply chain resilience has emerged as a critical priority following pandemic-related disruptions, with companies investing in visibility technologies, inventory optimization, and flexible sourcing strategies. Social commerce and influencer marketing are creating new pathways to consumers, particularly among younger demographics, while changing the dynamics of brand building and customer acquisition.

1.5.5 Financial Services Disruption and Innovation

The financial services industry continues to be reshaped by technological innovation, regulatory changes, and evolving customer expectations. Digital banking has become table stakes, with traditional institutions racing to modernize legacy systems while competing with agile fintech startups and technology giants expanding into financial services. The payments landscape is experiencing rapid innovation, with real-time payment systems, digital wallets, and cryptocurrency infrastructure challenging traditional transaction models. Open banking regulations and API standardization are enabling new ecosystems of financial services that transcend traditional institutional boundaries.

Artificial intelligence and machine learning are being deployed across functions from fraud detection to personalized wealth management, while also raising new questions about algorithmic bias and financial inclusion. Insurtech companies are leveraging IoT data, telematics, and advanced analytics to create usage-based insurance products and streamline claims processing. The emergence of decentralized finance and digital assets represents both a disruptive threat and strategic opportunity for traditional financial institutions. Meanwhile, cybersecurity and data privacy have become increasingly complex concerns as financial services become more digital and interconnected. Regulatory technology solutions are helping institutions manage compliance burdens while creating new business opportunities for specialized providers.

1.5.6 Industrial and Manufacturing Sector Trends

The industrial and manufacturing sector is in the midst of a technological revolution often described as Industry 4.0, characterized by the integration of digital technologies into production systems and business operations. Smart factories leveraging Internet of Things sensors, artificial intelligence, and advanced robotics are becoming more common, enabling unprecedented levels of automation, efficiency, and customization.

Additive manufacturing continues to evolve beyond prototyping to become a viable production method for specialized components, with applications expanding across aerospace, medical devices, and automotive industries. Industrial companies are developing new service-based business models centered around predictive maintenance, equipment-as-a-service, and performance optimization. Sustainability considerations are driving investment in energy efficiency, circular economy principles, and green manufacturing processes. Supply chain digitization is creating new opportunities for visibility, optimization, and collaboration across extended manufacturing networks. The convergence of information technology and operational technology is breaking down traditional silos but also creating new cybersecurity vulnerabilities that require specialized approaches. Workforce challenges persist as the skills required for advanced manufacturing evolve, forcing companies to invest in training, partnerships with educational institutions, and new approaches to talent attraction and retention.

1.5.7 Energy Sector Transition and Transformation

The energy sector is undergoing a historic transition driven by climate imperatives, technological advancements, and changing geopolitical dynamics. Renewable energy sources, particularly solar and wind, have achieved cost parity with traditional fossil fuels in many markets, accelerating the shift toward decarbonized energy systems. Electric vehicle adoption is progressing faster than many predictions, creating new challenges and opportunities for utilities, grid operators, and energy providers. Oil and gas companies are facing increasing pressure from investors, regulators, and consumers to articulate credible transition strategies that align with climate goals while maintaining energy security.

The hydrogen economy is moving from concept to early implementation, with significant investments in production, storage, and transportation infrastructure. Energy storage technologies are becoming increasingly critical for managing intermittency in renewable-dominated grids, with innovations in battery chemistry and alternative storage methods advancing rapidly. Digital technologies including artificial intelligence, advanced sensors, and blockchain are being deployed to optimize energy production, distribution, and consumption. The geopolitical landscape of energy is being reshaped by these transitions, with traditional energy relationships evolving and new dependencies emerging around critical minerals and manufacturing capabilities for clean energy technologies.

1.5.8 Regional Market Variations and Opportunities

While many industry trends display global characteristics, significant regional variations create a complex patchwork of opportunities and challenges for multinational organizations. North American markets continue to lead in technology adoption and digital transformation, with strong venture capital ecosystems supporting innovation across sectors. European markets are characterized by more stringent regulatory environments, particularly regarding data privacy, sustainability, and labor practices, while also benefiting from highly developed infrastructure and skilled workforces. The Asia-Pacific region represents the most dynamic growth market, with rapidly expanding middle classes, aggressive digital adoption, and strong government support for strategic industries.

Fig: Regional Differences Matter in Market Research

Emerging markets in Latin America, Africa, and Southeast Asia offer significant growth potential but present challenges related to infrastructure, political stability, and institutional capacity. Demographic trends vary dramatically, with aging populations creating particular challenges in Japan and Western Europe while younger populations in Africa and South Asia represent potential demographic dividends. Regional specialization is becoming more pronounced, with certain geographic areas developing concentrated expertise in specific technologies or industries. Trade relationships are evolving in response to geopolitical tensions and pandemic-related supply chain disruptions, with companies developing more regionalized manufacturing and distribution strategies to balance efficiency with resilience.

CHAPTER 2

Core Technologies in Electric Aviation

2.1 ELECTRIC PROPULSION SYSTEMS

2.1.1 Electric Motors: Types, Design and Efficiency

Electric motors are a mature technology that continues to evolve, shaped by materials science, power electronics, and clever control. Choosing among brushed, permanent magnet, induction, and reluctance families is only the opening bid; the real performance comes from the harmony of electromagnetic design, thermal management, inverter selection, and application-aware control.

Efficiency is not a single number but a landscape across operating conditions, and the best designs pay attention to every contributor—copper, iron, mechanical, and inverter losses—while integrating cooling and acoustics from the start. As the world electrifies transport and industry, motors will shoulder an even larger share of the energy bill. The winners will be those that align topology with duty cycle, leverage modern materials and SiC/GaN inverters, and close the loop with data to keep the system operating in its efficiency sweet spot, day after day.

Major Types of Electric Motors

Electric motors fall into a few foundational families, each optimized for particular performance envelopes, cost targets, and control strategies. Brushed DC motors are the historical starting point: simple, inexpensive, and easy to control with voltage, thanks to mechanical commutation using brushes and a commutator. They offer high starting torque and smooth low-speed behavior but demand regular maintenance due to brush wear, produce electrical noise, and struggle with efficiency at high speeds.

Brushless DC (BLDC) and permanent magnet synchronous motors (PMSM) replaced mechanical commutation with an electronic inverter, greatly boosting reliability and power density. The distinction between BLDC and PMSM is largely in back-EMF shape and control method: BLDC typically uses trapezoidal back-EMF and six-step commutation, while PMSM favors sinusoidal back-EMF with field-oriented control for smoother torque and lower acoustic noise. Both rely on high-energy permanent magnets (often NdFeB) to provide a constant rotor field, delivering excellent efficiency and compactness, especially in traction and servo applications. Induction motors (IMs), also called asynchronous motors, use a rotor with conductive bars (a "squirrel cage") and no brushes or magnets.

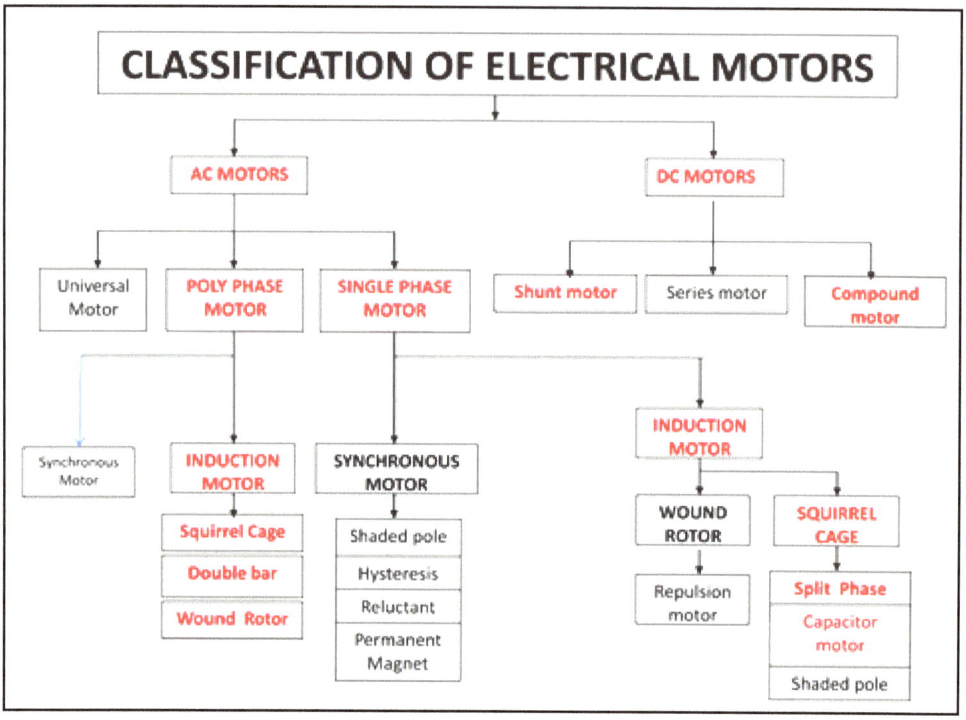

Fundamental Electric Motor Types

A rotating magnetic field in the stator induces currents in the rotor, producing torque with a characteristic slip between rotor speed and synchronous speed. Induction motors are rugged, low-cost, magnet-free, and ubiquitous in industry. Paired with variable frequency drives (VFDs), they achieve good efficiency across a broad speed range, though peak efficiency and torque density typically trail magnet-based machines.

Reluctance motors exploit the tendency of magnetic flux to follow the path of least reluctance. Switched reluctance motors (SRMs) have a simple, even tooth-like rotor with no windings or magnets, and a concentrated-winding stator that is sequentially energized. They are robust and can tolerate high temperatures and speeds, but exhibit torque ripple and acoustic noise unless carefully controlled. Synchronous reluctance motors (SynRMs), including interior permanent magnet variants (IPM-SynRM), use rotor saliency to produce torque with lower rare-earth content, offering a compelling compromise for traction where magnet reduction is a goal. Specialized motors serve precise niches. Stepper motors move in discrete steps with open-loop positional control, ideal for printers and light automation; microstepping improves smoothness at the cost of torque. Universal motors, which run on AC or DC, deliver very high speed and power-to-weight for tools and appliances but rely on brushes and are noisy. Geometry also matters: radial-flux motors are the classic cylindrical machines, while axial-flux designs deliver "pancake" form factors with high torque density and short end turns, increasingly attractive in EVs and aerospace. Linear motors unroll the stator and rotor to produce direct linear motion, useful in maglev and high-precision transport.

Core Design Principles

At the heart of every motor is a magnetic circuit comprising a laminated steel stator, a rotor, and copper windings. Torque arises from the interaction of stator-generated magnetic fields and rotor fields (from magnets, induced currents, or saliency). Key design levers include slot/pole count, winding configuration (distributed vs. concentrated), air-gap length, and material selection. Electrical steel laminations minimize eddy currents; thinner laminations and high-silicon or cobalt alloys reduce core losses at higher frequencies. Copper fill factor, conductor shape (round wire vs. hairpin), and end-turn length determine resistive losses and thermal behavior. Designers chase torque density by maximizing flux and current within thermal and mechanical limits, while managing saturation and ensuring sufficient mechanical margin against centrifugal forces at high speed.

Thermal management ties directly to performance and life. Cooling options range from natural and forced air to liquid jackets and direct oil-spray cooling that targets the end turns and stator teeth. Lower operating temperature reduces copper resistance (improving efficiency), preserves magnet coercivity, extends insulation life, and stabilizes bearing lubrication. Insulation classes and thermal interface designs are chosen so that hot spots stay within safe limits under worst-case duty cycles.

Control Strategies and Power Electronics

Inverters and control algorithms are as critical as the motor. Scalar control (V/f) for induction motors is simple and robust, but vector control—field-oriented control (FOC)—unlocks higher torque per ampere and better dynamics by independently regulating torque and flux. Direct torque control (DTC) offers fast response with a different philosophy.

PMSM and BLDC machines benefit from FOC for smooth torque and high efficiency, while BLDC six-step commutation remains viable for cost-sensitive designs. SRMs demand precise current profiling and rotor position sensing to tame torque ripple and noise. Sensorless control infers position from back-EMF or saliency but may struggle at low speeds; resolvers or encoders add cost but improve reliability. Power semiconductor choice shapes efficiency and size. Silicon IGBTs dominate medium-power inverters, while silicon carbide (SiC) MOSFETs and gallium nitride (GaN) transistors push switching frequencies higher, shrinking passive components and improving partial-load efficiency. The trade-off is that higher electrical frequencies can increase iron losses and require careful electromagnetic and thermal co-optimization.

Efficiency is Won and Lost

Motor-system efficiency is the sum of many small wins. Copper losses (I^2R) rise with current and temperature; designers minimize them with higher bus voltage (reducing current for a given power), high fill factor, and short end turns. Core losses blend hysteresis (proportional to frequency and material properties) and eddy currents (sensitive to lamination thickness, flux density, and frequency), demanding low-loss steels and careful flux management. Mechanical losses—bearing friction and windage—scale with speed and air density; balanced rotors, low-drag seals, and appropriate lubrication reduce these penalties.

Stray load losses arise from slot harmonics, leakage flux, and non-idealities in winding distribution. Mitigation includes skewing stator or rotor slots, optimizing slot/pole combinations, and refining tooth geometry. In the inverter, conduction and switching losses depend on device selection, gate drive, and switching frequency; dead-time and current ripple also impact motor copper and iron losses. A system-level view is essential: pushing switching frequency too high to shrink filters may inflate iron losses more than the inverter saves.

Efficiency maps summarize performance across torque and speed, revealing "islands" where the combined motor-inverter is most efficient. Control strategies such as flux weakening extend top speed in PMSM by counteracting the back-EMF, but do so at the cost of additional losses. Standards like NEMA MG1 and IEC 60034 guide efficiency measurement, with IE classes (IE1–IE5) and "premium" designations increasingly common for industrial motors. Because operating points vary, specifying efficiency at the duty-cycle level—considering speed profiles, load variability, and ambient conditions—is more meaningful than quoting a single peak value.

Applications and Selection Considerations

Industrial drives for pumps, fans, compressors, and conveyors account for the majority of motor electricity consumption. Here, variable speed control pays for itself quickly by matching torque to process demand, often delivering double-digit energy savings compared with throttling or fixed-speed operation. In electric vehicles, interior permanent magnet synchronous machines dominate for their torque density and efficiency, though induction and reluctance variants remain attractive where magnet cost, temperature robustness, or supply-chain risk rule. Household appliances have largely migrated to brushless designs for quietness and efficiency, while robotics and CNC systems rely on high-dynamic servo PMSMs and precise feedback devices. Selection hinges on duty cycle, environment, and total cost of ownership. High-torque, intermittent duty may favor magnet machines; harsh, high-temperature environments may push toward induction or reluctance. Packaging constraints (axial vs. radial dimensions), allowable noise and vibration, and available cooling all play decisive roles. Beyond motor choice, the integration quality—alignment, coupling, gearbox ratio, and inverter tuning—often makes or breaks system efficiency and reliability.

Design Trade-Offs and Constraints

Every motor is a compromise between torque density, efficiency, cost, manufacturability, and noise. Permanent magnets deliver superb performance, but add cost, demagnetization risk at high temperature, and exposure to rare-earth supply volatility. Magnet-free designs (induction, SRM, SynRM) reduce material risk and often cost, but require more current for the same torque, increasing copper and inverter losses, and can be harder to quiet acoustically. High speed shrinks the motor for a given power but puts greater stress on bearings and laminations and may require a gearbox, which introduces its own losses and reliability considerations. Electromagnetic design intersects with manufacturability. Fractional-slot concentrated windings simplify manufacturing and allow short end turns, yet can increase harmonic content and torque ripple unless carefully managed. Hairpin windings boost copper fill factor and improve thermal paths, but may suffer higher AC losses without proper transposition. Cooling choices ripple through materials, seals, and serviceability. As always, measure twice, cut once: prototypes and digital twins should iterate together to avoid late-stage surprises.

Emerging Trends and Innovations

Three trends stand out. First, tighter powertrain integration—motor, inverter, gearbox, and cooling in a single package—reduces cabling losses and improves thermal synergy, especially in EV e-axles and aerospace nacelles. Second, advances in materials and manufacturing are changing the game: axial-flux topologies, grain-oriented or thin-gauge steels, resin transfer molding, segmented stators, and additive manufacturing enable higher slot fill, shorter magnetic paths, and better cooling. Third, power electronics keep improving; SiC and GaN allow higher bus voltages and switching speeds, enabling smaller, lighter drives with excellent partial-load efficiency. On the sustainability front, magnet reduction and recycling are priorities. Designs that cut heavy rare-earth content via clever rotor topology, or that swap to ferrites with innovative flux paths, are moving from labs to pilot production. Digitalization rounds it out: physics-based digital twins and machine-learning diagnostics optimize control parameters in real time, detect incipient faults, and extend life through condition-based maintenance.

Ten Practical Ways to Improve Motor-System Efficiency

- Right-size the motor for the duty cycle, avoiding both chronic undersizing (overheating and high copper losses) and oversizing (low-load iron losses and poor inverter utilization).
- Use variable speed drives to match torque to process demand; even modest speed reductions on centrifugal loads yield disproportionately large energy savings.
- Increase bus voltage within safe limits to reduce current for a given power, cutting I^2R losses in both windings and power electronics.
- Specify low-loss materials—thin-gauge electrical steel, high-coercivity magnets, and high-fill windings—and design for short end turns to minimize parasitics.
- **Optimize Control:** Adopt field-oriented control, tune current regulators, and minimize dead-time and current ripple to shrink both torque ripple and losses.
- **Design for Cooling First, not Last:** Targeted liquid or oil-spray cooling lowers winding and magnet temperatures, preserving efficiency and extending life.
- Reduce mechanical losses through premium bearings, proper preload, low-drag seals, and careful rotor balancing to limit windage and vibration.
- Minimize gear ratio and transmission stages when possible; each stage adds losses and noise—direct drive or integrated gearsets can outperform add-on gearboxes.
- Validate with efficiency maps across torque and speed, not just peak points; use these maps to schedule operating strategies and control setpoints.
- Maintain the System: Alignment, lubrication intervals, filter cleanliness, and firmware updates for the drive can reclaim several percentage points of efficiency over time.

2.2 BATTERY TECHNOLOGIES

2.2.1 Lithium-Ion

Lithium-ion technology is the indispensable bridge to the electric aviation era. It is a mature, well-understood technology that has successfully moved from our pockets to our vehicles, and now, to our aircraft. While it possesses inherent limitations in energy density that currently restrict electric aircraft to short-range missions, and presents novel safety challenges that require heavy, complex mitigation systems, it remains the only viable option for current commercialization. The near-future of electric aviation will be defined by iterative improvements to this Li-ion baseline—optimizing chemistries, improving packaging efficiency, and enhancing safety—until next-generation energy storage solutions can fully mature to take its place.

Dominance of Lithium-ion Chemistry

Lithium-ion batteries dominate aviation applications because they currently offer the best commercially available compromise between energy density, power density, cycle life, and established manufacturing scale. The fundamental mechanism involves lithium ions moving from the negative electrode (anode) to the positive electrode (cathode) during discharge, and back again during charging. This "rocking chair" mechanism is highly efficient. In the context of aviation, every gram matters.

Li-ion batteries provide a much higher "specific energy" (energy per unit mass, measured in Wh/kg) than older rechargeable chemistries. While standard Li-ion cells might achieve 250-270 Wh/kg at the cell level, this is still orders of magnitude less than jet fuel. However, it is just enough to enable short-range missions, such as one-hour training flights or urban air taxi hops, which were previously impossible with heavier battery types.

Characteristics Defining Aviation Applications

The application of Li-ion batteries in aviation is far more complex than in ground vehicles. Aircraft require immense bursts of power for takeoff and vertical lift, necessitating batteries with high discharge rates (C-rates) that don't overheat instantly. Furthermore, they must be incredibly reliable; an electric car can pull over to the side of the road if a battery fails, but an aircraft cannot. This necessitates a deeper look at the specific characteristics and variations of Li-ion technology that are most relevant to aerospace engineering. The following ten aspects highlight the critical parameters and variations of lithium-ion usage in the skies:

- **NMC (Nickel Manganese Cobalt) Chemistry:** This is currently the most favored cathode chemistry for many aviation applications due to its high specific energy, allowing for the maximum possible range in the lightest package, though it can be more volatile than other types if damaged.
- **LFP (Lithium Iron Phosphate) Chemistry:** While offering lower energy density than NMC (making the aircraft heavier for the same range), LFP is gaining traction in some aviation sectors because it is inherently more stable, less prone to thermal runaway, and offers significantly longer cycle life.

- **Cell-Level vs. Pack-Level Density:** There is a critical distinction in aviation; while individual Li-ion cells may have high energy, once they are assembled into a flight-certifiable pack with necessary cooling systems, fire containment, and wiring, the effective energy density often drops by 30-40%.
- **Discharge Rate (C-Rate) Capability:** Aviation batteries must handle extreme C-rates during takeoff or hover phases. A high C-rate allows rapid energy release, but it generates significant internal heat that must be aggressively managed to prevent damage.
- **Thermal Runaway and Propagation:** The primary safety risk for Li-ion in aviation is thermal runaway, where one cell overheats and ignites its neighbors. Aviation standards require that a battery pack must contain any single-cell thermal runaway event without it propagating to the rest of the aircraft.
- **State of Charge (SOC) Operating Window:** To ensure safety and longevity, aviation Li-ion packs are rarely used from 100% down to 0%. They typically operate within a restricted buffer (e.g., 90% to 20%) to prevent stress on the chemistry, which further limits practical flight range.
- **Battery Management System (BMS) Criticality:** In aviation, the BMS is a flight-critical component. It must continuously monitor individual cell voltages, temperatures, and health, instantly isolating faults to prevent catastrophic failure during flight.
- **Cycle Life Economics:** The operational cost advantage of electric aircraft depends on how long Li-ion packs last. If frequent fast-charging for high utilization degrades the pack quickly, the high cost of pack replacement can negate the savings from not burning jet fuel.
- **Pre-Flight Thermal Conditioning:** Unlike cars, aviation batteries often need to be pre-heated or pre-cooled before flight to ensure they are in the optimal temperature range to deliver maximum power safely during the critical takeoff phase.
- **Silicon-Anode Enhancements**: To squeeze more performance out of standard Li-ion, emerging technologies are adding silicon to the graphite anode, which can significantly boost energy density, serving as a near-term bridge technology before solid-state batteries mature.

Safety and Certification Hurdle

The greatest challenge for Li-ion in aviation is not just performance, but certifiable safety. Aviation regulatory bodies like the FAA and EASA have extremely stringent requirements. The volatility of typical high-energy Li-ion liquid electrolytes presents a unique fire risk. Jet fuel is hard to ignite and generally only burns if there is an external ignition source. A Li-ion battery, however, can ignite internally if it suffers a manufacturing defect, internal short circuit, or physical damage. This leads to "thermal runaway," a self-sustaining fire that is incredibly difficult to extinguish and emits toxic gases.

For an electric aircraft to be certified, the manufacturer must prove that even if a worst-case thermal runaway occurs, it will be contained within the battery pack and allow the aircraft to land safely. This requirement adds significant non-energy-storing weight to the aircraft in the form of heavy containment vessels, robust cooling systems, and intumescent materials designed to block heat propagation. This "safety weight" directly counteracts the gains in battery energy density, creating a frustrating paradox for engineers trying to maximize flight range.

Operational and Economic Implications

The operating realities of Li-ion batteries also dictate the business models of early electric aviation. Because these batteries degrade faster when held at a high state of charge or when subjected to rapid charging, operators face a trade-off between high aircraft utilization and battery longevity. Air taxi services, which plan to fly many short hops per day with fast charging in between, may find their Li-ion packs degrade within a year or two, necessitating expensive replacements.

Conversely, flight schools that might charge more slowly overnight could see much longer pack lives. Furthermore, the "end of life" for an aviation battery is different from consumer electronics; once a pack degrades to perhaps 80% of its original capacity, it may no longer provide the necessary safety reserves for flight, even if it is still functional for less demanding ground applications (second-life use).

2.2.2 Solid-State

Solid-state electric aviation represents a significant leap forward in aerospace technology, offering the potential to revolutionize flight through higher efficiency, enhanced safety, and reduced environmental impact. By leveraging solid-state batteries and advanced electronic systems, electric aircraft can achieve greater energy density, lower operational costs, and near-zero emissions. While challenges remain—such as battery manufacturing, weight optimization, and thermal management—the trajectory of research and development is promising.

Hybrid-electric solutions provide transitional pathways, and continuous innovation in materials, power electronics, and energy storage will likely overcome current limitations. Aviation will increasingly rely on solid-state technologies to achieve sustainable, quiet, and efficient flight. As these systems mature, the skies may soon see electric aircraft operating alongside conventional planes, marking a pivotal moment in the history of aviation. Solid-state electric aviation is not just a technological evolution; it is a paradigm shift that has the potential to redefine the environmental and operational standards of air transport for decades to come.

Solid-State Technologies in Aviation

Solid-state technology refers to devices or systems that operate without moving liquids or mechanical components, relying instead on solid materials to store or conduct energy. In the context of electric aviation, the term often specifically refers to solid-state batteries (SSBs), which replace the liquid or gel electrolytes in conventional lithium-ion batteries with solid electrolytes. These batteries provide several advantages, including higher energy density, improved safety, and longer lifespans.

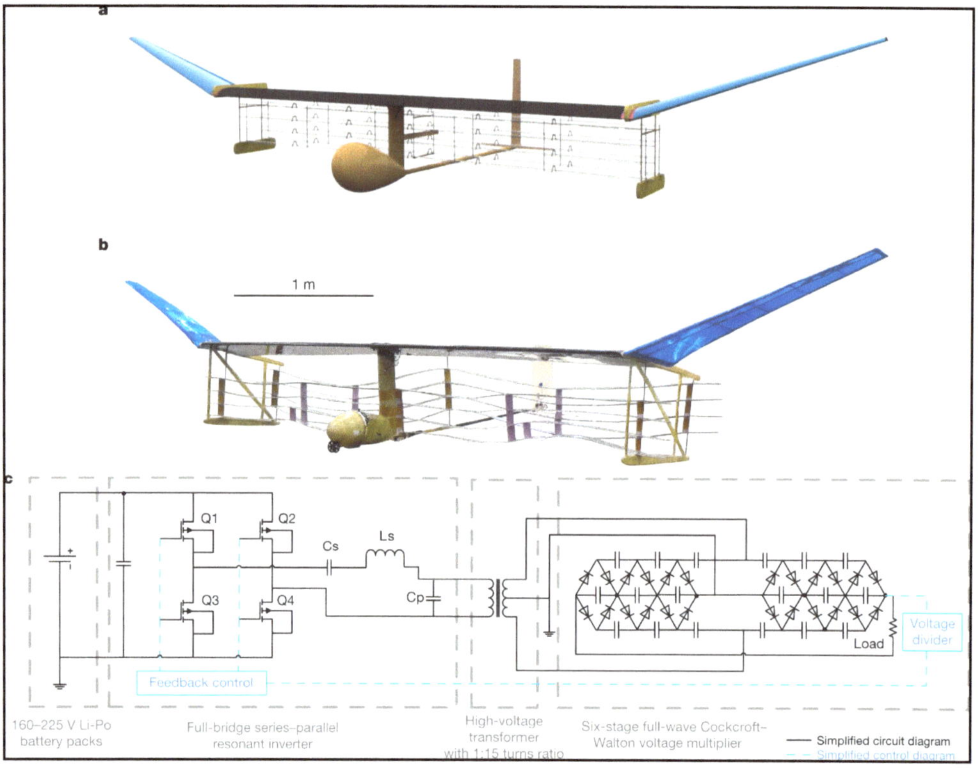

Integration of Solid-State Devices in Aviation

Beyond batteries, solid-state electronics in aviation include advanced power electronics, inverters, and motor controllers that manage electrical energy with high efficiency and minimal losses. The integration of these systems enables all-electric aircraft or hybrid-electric configurations capable of reducing reliance on fossil fuels while improving performance metrics such as thrust-to-weight ratio, range, and operational efficiency.

Advantages of Solid-State Batteries in Aviation

Solid-state batteries offer a range of advantages that make them ideal for electric aviation applications. First and foremost, their energy density is substantially higher than that of conventional lithium-ion batteries. This higher density allows aircraft to carry more energy for the same weight, extending flight range and payload capacity. In practical terms, this could mean that regional electric aircraft can perform longer missions without the need for frequent recharging. Another critical advantage is safety. Traditional lithium-ion batteries contain flammable liquid electrolytes, which can pose risks of thermal runaway and fire, particularly under the high-stress conditions of aviation. Solid-state batteries use non-flammable solid electrolytes, reducing the risk of catastrophic failure. Additionally, they operate efficiently across a broader temperature range, which is crucial for high-altitude flight where temperatures can drop significantly. Finally, solid-state batteries demonstrate longer cycle life and reduced degradation over time. This increases reliability and reduces maintenance costs, a major consideration for commercial aviation operators. The combination of energy density, safety, and longevity positions solid-state batteries as a cornerstone of sustainable electric aviation.

Electric Propulsion and Solid-State Integration

Electric propulsion systems rely heavily on the capabilities of energy storage and power management. Solid-state batteries provide a high-performance energy source that can supply consistent power to electric motors, enhancing thrust, acceleration, and operational efficiency.

Table: Comparison of Solid-State vs. Conventional Aviation Systems

Feature	Solid-State Electric Aviation	Conventional Aviation
Energy Storage	Solid-state batteries	Jet fuel or avgas
Energy Density	High (future 400–500 Wh/kg)	Moderate (≈43 MJ/kg for jet fuel)
Safety	Non-flammable, stable	Flammable, risk of fire
Efficiency	>90% motor efficiency	30–40% thermal efficiency
Noise	Very low	High engine and exhaust noise
Maintenance	Minimal, fewer moving parts	Complex, frequent servicing
Emissions	Near-zero	High CO_2 and NO_x emissions
Weight Impact	Battery weight significant, but shrinking	Fuel weight decreases as consumed
Lifespan	Long cycle life	Limited, requires regular overhauls
Operational Flexibility	Short-haul, urban, regional	Long-haul, high-speed operations

When combined with distributed electric propulsion (DEP), which uses multiple smaller motors along the wings or fuselage, solid-state systems can optimize lift-to-drag ratios, improve redundancy, and enable innovative aircraft designs. Electric motors powered by solid-state systems are inherently more efficient than conventional combustion engines. Typical electric motor efficiency exceeds 90%, compared to 30–40% for traditional gas turbines. This high efficiency translates into lower energy consumption per flight hour, reduced emissions, and quieter operation. Aircraft such as the Eviation Alice and prototypes from Pipistrel have demonstrated the feasibility of all-electric flight, and the next generation of aircraft will increasingly rely on solid-state energy solutions to enhance performance and safety.

Technological Challenges in Solid-State Electric Aviation

Despite its advantages, solid-state electric aviation faces several technological challenges. The primary concern is battery manufacturing scalability. Producing solid-state batteries at the size and volume required for commercial aircraft remains complex and costly. Material availability, manufacturing precision, and electrolyte stability are critical factors that must be addressed before widespread adoption. Another challenge is thermal management. Although solid-state batteries are safer than conventional lithium-ion systems, high-power operations generate heat that must be effectively dissipated to maintain efficiency and prevent performance degradation. Advanced cooling systems and thermal designs are therefore essential in aircraft engineering. Weight and energy density trade-offs also present obstacles. While solid-state batteries promise higher energy density, they are still heavier than conventional fuel sources when considering current technology. Engineers must optimize aircraft structures, aerodynamic efficiency, and energy management systems to ensure practical flight ranges and payload capacities.

Hybrid Approaches and Transitional Solutions

Given the current limitations of battery technology, hybrid-electric aircraft are emerging as a transitional solution. Hybrid configurations combine conventional combustion engines with electric propulsion powered by solid-state batteries. These aircraft can optimize fuel efficiency, reduce emissions, and extend flight range beyond what battery-only systems can currently achieve.

Hybrid approaches also allow testing and gradual integration of solid-state systems without fully committing to unproven technology. Examples include hybrid regional aircraft prototypes, which use electric motors for taxiing and low-speed flight while relying on conventional engines for long-distance cruise. This approach not only enhances efficiency but also reduces noise and fuel consumption in airports and urban environments.

Environmental and Sustainability Benefits

Solid-state electric aviation promises profound environmental benefits. By reducing or eliminating reliance on fossil fuels, these systems can dramatically cut greenhouse gas emissions and particulate pollution. Additionally, the low operational noise of electric motors makes electric aircraft well-suited for urban air mobility (UAM) and short-haul regional flights, where noise pollution is a major concern.

The sustainability of solid-state aviation also depends on the source of electricity. Charging batteries with renewable energy sources such as solar, wind, or hydroelectric power ensures that the environmental footprint remains minimal. Furthermore, advances in battery recycling and material reuse can mitigate the ecological impact of battery production and disposal.

Solid-State Electric Aviation

- **High Energy Density:** Enables longer flights with lighter battery packs.
- **Enhanced Safety:** Solid electrolytes reduce fire and thermal runaway risks.
- **Long Cycle Life:** Batteries retain capacity over many charge-discharge cycles.
- **High Efficiency:** Electric motors convert most stored energy into thrust.
- **Low Noise:** Suitable for urban and regional operations.
- **Reduced Maintenance:** Fewer moving parts compared to combustion engines.
- **Near-Zero Emissions:** Supports sustainable aviation initiatives.
- **Scalability Challenges:** Manufacturing and material constraints exist.
- **Thermal Management Requirements:** High-power operations produce heat that must be controlled.
- **Hybrid Integration Potential:** Can complement conventional propulsion in transitional aircraft designs.

These features highlight the transformative potential of solid-state systems while acknowledging the current technological hurdles that must be overcome.

Applications and Future Prospects

Solid-state electric aviation is particularly suited for short-haul regional aircraft, urban air mobility vehicles, and training aircraft, where flight distances are manageable, and efficiency, safety, and noise reduction are priorities. Small electric commuter planes and air taxis are likely to be the first commercial applications, benefiting from the high torque, silent operation, and reduced environmental impact of electric propulsion.

In the future, as solid-state battery technology matures, we can expect larger aircraft with longer range capabilities. Continuous research into energy-dense solid-state chemistries, advanced power electronics, and lightweight materials will expand the feasibility of fully electric or hybrid-electric commercial aircraft. Additionally, regulatory frameworks and certification standards are evolving to accommodate the unique characteristics of solid-state electric aviation, paving the way for safe and widespread adoption.

Economic Considerations

While initial acquisition costs of solid-state electric aircraft are high due to battery technology and integration complexities, operational costs are expected to be lower than conventional aircraft. Electricity is cheaper and more stable in price than aviation fuel, and reduced maintenance needs further decrease total lifecycle costs. Airlines may also benefit from government incentives for low-emission aircraft, particularly in regions with strict environmental regulations. Over time, as solid-state battery production scales and costs decline, electric aviation could become economically competitive with conventional propulsion. Moreover, the lower operating costs and environmental benefits may provide a competitive advantage for airlines adopting early electric fleets.

2.2.3 Metal-Air

Metal-air battery technology stands at a pivotal juncture, holding the key to unlocking a future of clean, efficient, and practical electric aviation. While formidable challenges related to rechargeability, lifespan, and system integration remain, the pace of innovation is accelerating. The fundamental physics of metal-air systems offers a pathway to energy densities that could enable meaningful electric flight, potentially transforming regional air travel and significantly reducing aviation's environmental impact.

Through global collaboration between academia, startups, and industry, the goal of powering regional aircraft with safe, high-performance metal-air batteries is moving from the realm of theory toward tangible reality. The journey to commercialization will require sustained investment and interdisciplinary collaboration across materials science, electrochemistry, and aerospace engineering. Success will likely come in phases, beginning with hybrid implementations and progressing toward more fully metal-air powered aircraft as the technology matures. The potential reward is substantial: a genuinely sustainable aviation system that maintains the connectivity benefits of air travel while dramatically reducing its environmental footprint. As these advancements mature, metal-air batteries are poised to play a critical role in decarbonizing the aviation sector, revolutionizing how we fly and creating a cleaner future for air transportation.

Operational Principle and Key Chemistries of Metal-Air Batteries

At its core, a metal-air battery generates electricity through the electrochemical reaction between a metal and oxygen. The basic components include a metal anode, an air cathode with a catalyst to facilitate the oxygen reaction, and an electrolyte that allows ions to move between the electrodes. During discharge, the metal anode oxidizes, releasing electrons that travel through an external circuit to power a device.

Simultaneously, oxygen from the atmosphere diffuses into the porous cathode, where it combines with the electrons and ions from the electrolyte to form a metal oxide or hydroxide as a byproduct. The specific theoretical energy output varies dramatically depending on the metal used. Several metal-air chemistries show particular promise for aviation applications. Lithium-air batteries offer the highest theoretical energy density of all metal-air systems, potentially exceeding 11,000 Wh/kg.

However, they face significant challenges with rechargeability and stability. Aluminum-air batteries provide a compelling balance of high energy density and material abundance, making them attractive for cost-sensitive applications. Zinc-air batteries represent a more mature technology with established manufacturing processes and good safety characteristics. Sodium-air batteries offer potential cost advantages while still delivering impressive energy density figures. Each chemistry presents unique trade-offs between energy density, power density, cycle life, and material costs that must be carefully considered for aviation applications.

Advantages of Metal-Air Batteries for Aviation Applications

The unique chemistry of metal-air batteries confers several compelling advantages for aviation, with energy density being the most significant. The theoretical energy densities of metals like lithium and aluminum far surpass not only current lithium-ion batteries but also the 1,000 Wh/kg target for electric aviation.

This could enable regional flights of up to 300 miles, covering a substantial portion of domestic air travel. A second major advantage is the potential for lower costs and greater sustainability. Many of the metals used, particularly zinc and aluminum, are more abundant and less expensive than lithium. Furthermore, their environmental profile is often more favorable; aluminum-air batteries, for instance, can be designed to use post-consumer recycled aluminum scrap as fuel, creating a circular economy model.

- **Exceptional Energy Density**: Metal-air batteries theoretically offer 5-10 times the energy density of conventional lithium-ion batteries, making them uniquely suited for weight-sensitive aviation applications.
- **Material Abundance**: Key metals like aluminum, zinc, and sodium are among the most abundant elements on Earth, providing a sustainable alternative to scarce materials.
- **Reduced Environmental Impact**: The operational emissions of metal-air batteries are zero, and their components can often be recycled more efficiently than conventional battery materials.
- **Safety Advantages**: Certain metal-air systems offer enhanced safety characteristics due to the physical separation of reactants and the use of aqueous electrolytes.
- **Simplified Thermal Management**: The operating principles of metal-air batteries can reduce the complexity of thermal management systems compared to high-density lithium-ion packs.
- **Potential for Rapid Refueling**: Mechanically rechargeable metal-air systems could be refueled by replacing the metal anode, potentially faster than recharging conventional batteries.
- **Scalability**: The modular nature of metal-air cells allows for flexible pack designs that can be optimized for different aircraft configurations.
- **Reduced Resource Geopolitics**: By utilizing widely available metals, metal-air batteries could reduce dependence on geographically concentrated resources like cobalt and lithium.
- **Compatibility with Hybrid Systems**: Metal-air batteries can be effectively paired with other power sources in hybrid-electric aircraft architectures.
- **Infrastructure Synergies**: The metals used in these batteries often have established supply chains and recycling infrastructure that can be adapted for aviation use.

Critical Challenges and Ongoing Research

Despite their immense promise, metal-air batteries face significant scientific and engineering hurdles that must be overcome before they can be certified for aviation use. A primary challenge is their limited cycle life and rechargeability. Many metal-air chemistries, particularly those using aluminum or magnesium, are not yet practically rechargeable in the conventional sense. Instead, they operate as mechanically rechargeable systems, where the spent metal anode is physically replaced with a fresh one. Researchers are actively developing advanced catalysts and stable electrolytes to create truly rechargeable systems, but achieving long cycle life remains a key focus. Another major hurdle is managing parasitic reactions and component degradation. The air cathode can become clogged by reaction byproducts, and the metal anode may corrode when in contact with the electrolyte, leading to self-discharge and reduced efficiency. Sensitivity to environmental factors is also a concern; the performance of some batteries can be degraded by carbon dioxide and humidity in the air, requiring sophisticated air management systems. Finally, the recycling infrastructure for many metal-air battery types is still underdeveloped compared to that for lead-acid or lithium-ion batteries, posing a challenge for their end-of-life sustainability. Research institutions and companies worldwide are tackling these challenges through nanomaterials engineering, advanced electrolyte formulations, and innovative cell designs that could unlock the full potential of metal-air technology for aviation.

Competitive Landscape: Startups and Innovations

The race to commercialize metal-air technology for high-impact applications like aviation is being driven by a dynamic ecosystem of startups and research institutions, each tackling the core challenges from different angles. Companies are developing innovative approaches to overcome the traditional limitations of metal-air systems. Some focus on novel electrolyte compositions that prevent electrode degradation while maintaining high ionic conductivity. Others are engineering sophisticated cathode structures that maximize oxygen diffusion while minimizing contamination. Several ventures are exploring hybrid systems that combine metal-air chemistry with conventional batteries to balance energy density with power delivery. Academic institutions play a crucial role in advancing the fundamental science behind metal-air batteries. Research groups are investigating new catalyst materials that can efficiently drive the oxygen reduction and evolution reactions without precious metals. Others are developing protective coatings for metal anodes that suppress unwanted side reactions. The interface between academic research and industrial application is particularly active, with several university spinouts working to translate laboratory breakthroughs into commercially viable products. Established aerospace companies are also investing in metal-air research through partnerships and internal development programs, recognizing the transformative potential of this technology for electric aviation.

Integration Challenges in Aviation Applications

Adapting metal-air battery technology for aviation use presents unique integration challenges that extend beyond the core electrochemical performance. The weight advantage of metal-air systems must be balanced against the additional components required for their operation in an aircraft environment. Air management systems must be designed to supply sufficient oxygen for the electrochemical reaction while filtering out contaminants like carbon dioxide and water vapor that can degrade performance. Thermal management presents another significant challenge, as the reaction kinetics and efficiency of metal-air cells are temperature-dependent, requiring careful thermal system design. Safety certification represents a particularly complex hurdle for metal-air batteries in aviation.

The use of reactive metal powders or foils combined with air-breathing architecture creates failure modes that differ significantly from conventional batteries. Comprehensive testing must demonstrate that metal-air systems can withstand aviation's rigorous environmental conditions, including pressure changes, vibrations, and temperature extremes, without compromising safety. The development of approved maintenance procedures for mechanically rechargeable systems adds another layer of complexity. Furthermore, the power characteristics of metal-air batteries may require complementary energy storage systems for high-power maneuvers like takeoff and climbing, leading to hybrid architectures that increase system complexity but may offer the most practical near-term pathway to implementation.

Comparative Analysis with other Emerging Technologies

Metal-air batteries exist within a broader ecosystem of emerging energy storage technologies competing to enable electric aviation. When compared to advanced lithium-ion batteries, metal-air systems typically offer superior energy density but lag in power density and cycle life. Against hydrogen fuel cells, metal-air batteries may provide advantages in volumetric energy density and infrastructure compatibility, though fuel cells currently achieve faster refueling times.

The comparison with sustainable aviation fuels reveals a different set of trade-offs, with metal-air batteries offering zero operational emissions but facing energy density limitations compared to liquid fuels. The most promising near-term applications may involve hybrid systems that leverage the strengths of multiple technologies. A regional aircraft might utilize high-power lithium-ion batteries for takeoff and climb, complemented by metal-air batteries for cruise flight where energy density is most critical. Alternatively, metal-air systems could be paired with hydrogen fuel cells in a hybrid configuration that optimizes both range and power availability.

The ultimate suitability of each technology will depend on specific aircraft missions, with metal-air batteries showing particular promise for short-haul regional routes where their energy density advantage can be fully leveraged without requiring excessive cycle life. As all these technologies continue to evolve, the competitive landscape will shift, potentially creating opportunities for metal-air batteries to capture specific market segments where their unique characteristics provide decisive advantages.

2.2.4 Energy Density

Electric aviation's future will be decided less by a moonshot chemistry and more by accumulated, compounding gains. A 10% improvement in pack-level energy, a 5% gain in aerodynamic efficiency, a 3% reduction in drivetrain losses, and smarter operations together can turn a marginal mission into a profitable one. In the near term, the strongest use cases will be those that respect energy's constraints: eVTOL routes that minimize hover and exploit high-frequency corridors; regional fixed-wing services over 100–300 km with high L/D airframes and reliable charging; and cargo or medical logistics where payload is modest and time certainty is paramount.

Hybrids—pairing batteries with efficient turbogenerators—can bridge gaps for longer routes while battery technology matures. As cells inch upward in density and certification de-risks new pack architectures, the effective energy density available to operators will rise faster than the chemistry alone suggests. Better thermal systems, structural integration, high-voltage distribution, and software that flies the battery as thoughtfully as the airframe will unlock longer ranges and higher payloads without waiting for a physics revolution. The sky won't go fully electric overnight, but for short-haul missions, the climb is already underway—and energy density, properly understood and engineered, is the ladder.

Energy Density Really Means

In aviation, energy density has two faces. Gravimetric energy density (Wh/kg) determines how much energy you can carry for a given weight—critical in aircraft, where every kilogram displaces payload or range. Volumetric energy density (Wh/L) matters when space is tight, influencing aerodynamics and packaging. Batteries also have two levels of density: cell-level, which chemistry determines in the lab, and pack-level, which is what the airplane actually flies with after adding structure, cooling, wiring, sensing, and safety systems. The gap between cell and pack can be 20–40%, so pack-level values drive mission planning. Finally, "effective" energy density reflects what is usable in operation after subtracting reserves, accounting for power-limited phases like takeoff and hover, and derating for temperature and aging.

Jet Fuel Versus Batteries: A Fair Comparison

Jet fuel's lower heating value is about 43 MJ/kg, roughly 12,000 Wh/kg. That looks unbeatable against today's batteries, but propulsion efficiency alters the comparison. Small turbines and piston engines convert only about 25–40% of that chemical energy into shaft power at the propeller. Electric drivetrains (inverter plus motor) routinely deliver 85–95% efficiency in cruise. If you translate energy into delivered shaft work, a kilogram of jet fuel still wins but by an order of magnitude, not two: think several thousand Wh/kg of shaft work for fuel versus a few hundred

Wh/kg for a battery pack. That gap shapes design choices—shorter routes, higher lift-to-drag, and, for vertical-lift aircraft, wings to escape hover early. As battery technology improves and operations optimize around electric strengths—short stage lengths, high utilization, precise energy management—the range at which electric becomes practical expands.

Pack-Level Reality in Aviation

Cell makers regularly report 300–400 Wh/kg at the cell level for high-nickel chemistries with silicon-doped anodes. Once integrated into an airworthy pack—with crashworthy enclosures, structural frames, liquid cooling, contactors, fuses, sensors, and fire containment—the pack-level gravimetric energy density typically lands much lower.

For near-term aviation-grade packs, 180–260 Wh/kg is a realistic range, with volumetric energy density around 400–700 Wh/L depending on cooling and packaging. Designs optimized for certification robustness tend to trade a few percent of density for thermal and mechanical margin. Advanced packs targeting 250–300 Wh/kg are plausible over the next few years as manufacturing, materials, and thermal designs mature, but the aviation certification cycle means lab breakthroughs often take a decade to reach service.

Mission Planning and Regulatory Reserves

Usable energy is not the full state-of-charge window. Operators hold energy reserves for contingencies, alternates, and loiter or hover time, and they maintain conservative guard bands to protect cycle life. Regulators are converging on reserve requirements for electric vertical and regional aircraft that, in practice, often translate into 20–30% of nominal capacity being unavailable for planned mission energy. Cold weather, which increases internal resistance, can further reduce available energy unless packs are preconditioned. Missions budget energy for taxi, takeoff, climb, cruise, descent, a missed approach or diversion, and a fixed reserve slice. The net result: a battery rated at 250 Wh/kg on paper may deliver closer to 150–180 Wh/kg of mission-usable energy once reserves and power-limited phases are accounted for.

Payload–Range: The Electric Breguet

Conventional aircraft range is often analyzed through the Breguet equation, where energy content, efficiency, and aerodynamic quality (lift-to-drag, L/D) set the envelope. Electric aircraft have a different, simpler constraint: the mission ends when you hit the energy floor. That foregrounds L/D and weight fraction. Fixed-wing electrics with L/D values of 12–18 can achieve meaningful regional ranges with today's packs, especially if they avoid step climbs and optimize cruise altitude for power draw. eVTOLs, by contrast, spend valuable minutes in hover or powered low-speed flight, where power per unit lift is highest. Architectures that minimize hover time—lift-plus-cruise with large rotors (low disk loading) and efficient wings—stretch range significantly. The golden rule is to transition to wing-borne flight as early as safe operations allow, protect cruise efficiency with clean aerodynamics, and land with adequate reserve.

Thermal and Power Limits that Shrink Usable Energy

Batteries deliver their best energy density at moderate discharge rates and temperatures. Aviation loads are spiky: takeoff and go-around demand short-duration high C-rates, while eVTOL hover can require sustained 3–5C with peaks higher. At high C-rate, voltage sag and ohmic heating reduce usable capacity; thermal management must cap temperature rise to protect safety and life. That means some missions are power-limited before they are energy-limited. Designers address this with larger packs (adding mass), parallel strings for lower current per cell (adding volume and complexity), and aggressive cooling. Each mitigation erodes the effective energy density at the aircraft level. Charging is similar: fast turnarounds require high charge power, but repeated fast charging accelerates aging and compels stricter thermal controls, again nudging system mass upward.

Chemistry Pathways and Timelines

Today's aviation-focused batteries mostly use high-nickel NMC or NCA chemistries with silicon-enhanced anodes to push energy higher while retaining acceptable cycle life. Lithium iron phosphate (LFP) offers excellent safety and cycle durability but trades away energy density, making it more suitable for training aircraft or shorter-range missions. Solid-state lithium-metal promises higher cell-level energy density (potentially 350–450 Wh/kg), lower flammability risk, and fast-charge capability, but scaling, manufacturability, and cycling at high C-rates remain the gating factors. Lithium–sulfur could deliver very high gravimetric energy with simpler materials, yet it struggles with cycle life and volumetric density. Sodium-ion's lower cost and good cold performance are attractive, but its density is presently too low for most aviation use. Realistically, the late 2020s and early 2030s will be about squeezing more from advanced Li-ion variants while piloting early solid-state packs through certification on limited missions.

Structural Integration and Airframe Design

Energy density is not just chemistry; it's architecture. Structural battery packs—where the enclosure and even some cell modules carry primary loads—can shave kilograms and improve crashworthiness by spreading loads. Integrating packs into wings or the fuselage floor helps center of gravity control and shortens cable runs, reducing copper mass and inverter losses. High-voltage systems (800–1000 V) cut current and cable size, improving both efficiency and packaging. Airframe design responds in kind: high aspect-ratio wings reduce induced drag; distributed electric propulsion can tailor lift distribution and reduce wing area; and carefully placed nacelles minimize interference drag. Each percent gain in aerodynamic or electrical efficiency translates directly to either more payload for the same battery or more range for the same mass—an "effective energy density" dividend that compounds.

Infrastructure and Operations that Multiply Energy

Ground operations influence airborne energy. Preconditioning batteries to the optimal temperature before departure recovers usable capacity and power. Smart charging strategies that soak at moderate rates and schedule fast-charge bursts sparingly preserve cycle life without compromising the timetable. Battery swapping can decouple utilization from charging, but it requires heavier, modular packs, robust connectors, and precise alignment—all mass and complexity penalties that chip away at density.

At vertiports and regional airports, energy storage buffers can absorb grid peaks, enabling consistent charge power and lower electricity cost. Route design matters too: shorter sector lengths with high frequency and predictable load factors allow smaller reserves and tighter state-of-charge windows than infrequent long routes with uncertain winds.

Safety and Certification Shape the Pack

Regulatory frameworks for battery safety (for example, standards focused on thermal runaway containment, venting, and fire detection) drive pack architecture. Segmented modules with robust isolation, high-speed contactors, gas and smoke sensing, and dedicated overpressure vent paths are the norm in certifiable designs. Designers choose conservative charge windows—rarely using 0–100%—to limit plating and extend life. These choices add mass and reduce nominal capacity, but they are non-negotiable for passenger safety. The art is to design cooling and containment so that a single-cell event is isolated and non-propagating, while preserving maintainability and minimizing mass penalties.

Economics of Energy Density

Every extra Wh/kg converts to either more seats sold permission or more missions per day. Higher energy density reduces the number of charge cycles per block hour, slows degradation, and cuts operating cost by spreading pack depreciation over more revenue flights. But energy density has a price: high-nickel cells and exotic separators cost more, and aggressive designs may shorten life if not carefully managed. The economic optimum is rarely the absolute highest density; it's the pack and mission pairing that yields the lowest cost per seat-mile while hitting safety and schedule targets. Operators will tune state-of-charge windows, charging rates, and route structures to keep the battery in its most economical zone.

Ten Levers to Boost Effective Energy Density

- Choose missions that minimize high C-rate demand, especially hover time; transitioning quickly to wing-borne flight preserves usable capacity and limits heat.
- Optimize aerodynamics for high lift-to-drag ratios and low disk loading; every percent drop in drag adds directly to "effective" energy per kilogram.
- Raise bus voltage within safety limits to reduce current and cable mass; thinner harnesses and lower I^2R losses improve both weight and efficiency.
- Integrate thermal management early: liquid jackets, oil spray, and preconditioning reclaim capacity in cold and prevent derates in heat.
- Design packs for maintainable modularity with minimal overhead; structural enclosures that carry load reduce redundant mass.
- Use advanced cells only where they pay back; pairing high-energy cells for cruise with high-power modules for takeoff avoids oversizing the whole pack.
- Protect cycle life with smart charging and SOC windows; modest charge rates and narrow operating bands maintain capacity over thousands of cycles.
- Instrument for data-driven operations; cell-level telemetry and predictive analytics enable tighter reserves without compromising safety.
- Align routes and schedules with local energy infrastructure; reliable charge power and off-peak pricing reduce overbuilding buffers that add mass.
- Continually refine the powertrain: low-loss inverters, efficient motors, and precise control can "add" tens of Wh/kg by cutting system losses.

2.2.5 Safety

Safety in electric aviation represents a complex merging of established aerospace rigor with novel electromechanical challenges. While the inherent simplicity of electric motors offers a promising path toward greater long-term reliability, the immediate hurdle involves mastering the volatile nature of high-energy batteries and high-voltage architectures. By addressing these unique risks through robust containment strategies, advanced monitoring systems, and comprehensive new operational protocols, the industry aims to deliver a new generation of aircraft that are not only environmentally sustainable but also maintain the impeccably high safety standards that the traveling public demands.

Primary Challenge: High-Energy Battery Safety

The most significant safety challenge in electric aviation centers on the energy storage medium, primarily lithium-ion batteries. Unlike jet fuel, which is relatively inert until it is atomized and ignited in a specific environment, a fully charged battery represents a tremendous amount of potential energy that is inherently "live."

The critical safety risk here is thermal runaway—a chain reaction where a single cell overheats due to internal failure, physical damage, or overcharging, and subsequently ignites adjacent cells. In traditional aviation, a fuel leak can often be managed by shutting off a valve. In electric aviation, a battery pack in thermal runaway cannot simply be "turned off" because the oxidizer required for the fire is contained within the battery chemistry itself. Therefore, safety in design focuses heavily on containment and isolation, ensuring that even if a worst-case runaway event occurs, it remains localized within a specific module and does not compromise the structural integrity of the airframe or allow toxic gases to enter the cabin.

High-Voltage Risks and Environmental Factors

To generate sufficient power for flight without incurring prohibitive weight penalties from thick copper cabling, electric aircraft must operate at very high voltages, often between 500V and 1,000V DC or higher. This introduces electrical safety risks previously reserved for industrial ground applications. At standard sea-level pressure, air is a decent electrical insulator. However, as an aircraft climbs and air density decrease, its insulating properties diminish significantly—a phenomenon described by Paschen's Law. This increases the risk of electrical arcing or corona discharge between high-voltage components if they are not adequately shielded. Such arcing can lead to electrical fires or complete loss of propulsion. Consequently, safety engineering in electric aviation involves developing specialized, highly insulated, and shielded wiring harnesses and connectors that can maintain their integrity across varied pressure and temperature regimes, safeguarding the aircraft against catastrophic electrical faults.

Critical Safety Considerations in Electric Design

The holistic safety approach for electric aircraft must address a disparate array of new vectors. It is not enough to just secure the battery; the entire ecosystem of the aircraft, from its software to its physical structure, must be optimized for this new form of propulsion. The industry is currently standardizing the following ten critical safety considerations that define the airworthiness of these new machines:

- **Thermal Runaway Containment Systems:** Designing battery enclosures with robust venting and fireproof materials that can withstand the intense heat and pressure of a multi-cell failure without breaching the aircraft structure.

- **High-Voltage Arc Suppression:** Utilizing advanced insulation materials and geometric spacing in electrical components to prevent arcing at high operational altitudes where air provides less natural insulation.
- **Battery Management System (BMS) Authority:** Developing flight-critical grade BMS software that can instantly detect minute anomalies in cell voltage or temperature and isolate problem areas before they cascade into failures.
- **Electromagnetic Interference (EMI) Shielding:** Ensuring the massive electromagnetic fields generated by high-power electric motors do not interfere with sensitive avionics, navigation equipment, or fly-by-wire control systems.
- **Crashworthiness of Heavy Mass Concentrations:** Reinforcing airframes to safely secure heavy battery packs during high-impact events, preventing them from breaking loose and crushing occupants during a survivable crash.
- **Reserve Energy Quantification:** Redefining standard "fuel reserve" regulations to account for battery behavior, ensuring aircraft have guaranteed landing power even after a maximum-range flight, despite varying battery health or temperature conditions.
- **Acoustic Safety Hazards:** Addressing the counter-intuitive risk of aircraft being too quiet on the ground, requiring artificial noise generation to warn ground crews and pedestrians of an active, live aircraft.
- **Ground Crew High-Voltage Training:** Establishing rigorous new safety protocols for maintenance personnel who must work on energized aircraft, where a single mistake with a tool could result in lethal electrocution.
- **Emergency Rescue Information:** Providing standardized, instantly accessible information for first responders so they know how to safely approach a crashed electric aircraft without cutting into high-voltage lines or unstable battery packs.
- **Distributed Propulsion Redundancy:** Leveraging the ability to use many small motors (eVTOLs often have 6 to 12) so that the complete failure of one or more propulsion units is a manageable, non-catastrophic event.

Operational Safety and Human Factors

Beyond the hardware, the human element of safety is undergoing a significant transition. Pilots trained on combustion engines rely on auditory and vibratory cues to sense the health of their aircraft. Electric motors are exceptionally smooth and quiet, potentially masking developing issues until they become critical.

Furthermore, failure modes in electric aircraft can be instantaneous. While a failing piston engine might cough and sputter, giving a pilot time to react, an electrical inverter failure can mean an instant total loss of thrust from that unit. Operational safety, therefore, depends on new pilot training curriculums that emphasize deep systems knowledge, enabling pilots to interpret highly digitized cockpit displays and react appropriately to novel emergency scenarios, such as battery over-temperature warnings or electrical isolation events.

Regulatory Safety Net

Ultimately, the safety of electric aviation is guaranteed by rigorous certification processes overseen by bodies like the FAA (USA) and EASA (Europe). Because current regulations (like Part 23 for small aircraft) were written for combustion vehicles, regulators are using "Special Conditions" to bridge the gap. These are bespoke sets of safety requirements tailored to the specific architecture of each new electric aircraft.

This process is deliberately slow and exacting. Manufacturers must prove, through destructive testing and millions of simulated flight hours, that their novel systems meet or exceed the 10^{-9} (one failure per billion flight hours) safety standard required for critical commercial aircraft systems. This regulatory conservatism, while frustrating to investors eager for rapid deployment, is the essential firewall that ensures electric aviation does not compromise the hard-won safety reputation of the broader industry.

2.2.6 Lifecycle Challenges

Electric aviation represents a transformative shift in aerospace technology, but it is accompanied by unique lifecycle challenges that span every phase of the aircraft's life. Design, material, operational, maintenance, environmental, regulatory, infrastructure, technological, economic, and market considerations all play critical roles in determining the feasibility and sustainability of electric aircraft.

These challenges and implementing targeted strategies—such as advanced battery development, hybrid propulsion, infrastructure investments, and regulatory frameworks—the aviation industry can overcome current limitations. Addressing lifecycle challenges is essential not only for operational success but also for realizing the environmental and economic benefits of electric aviation. The evolution of electric aircraft is not merely a technological shift; it represents a fundamental transformation in how society approaches sustainable air transportation, promising a future of cleaner, quieter, and more efficient flight.

Design and Development Challenges

The design phase of electric aircraft involves complex trade-offs between energy storage, weight, aerodynamics, and performance. Unlike conventional aircraft that burn fuel, electric aircraft rely on batteries or hybrid systems, which are heavier for the same amount of stored energy. This weight penalty directly affects payload capacity, range, and flight efficiency. Designers must carefully balance battery weight against aircraft structure and aerodynamics to optimize performance. Battery placement, cooling requirements, and integration with propulsion systems further complicate design. The aircraft's center of gravity changes as batteries discharge, affecting stability and control. Additionally, ensuring safety in the event of battery failure or thermal runaway requires advanced engineering solutions, including fire-resistant materials and robust battery management systems. These complexities increase design time and cost, posing significant challenges to widespread adoption.

Manufacturing and Material Constraints

Manufacturing electric aircraft involves materials and components that are not yet produced at large scales for aviation applications. High-energy-density batteries, solid-state or lithium-ion, require precise assembly and high-quality materials. Supply chain constraints for lithium, cobalt, nickel, and other critical materials present risks for scalability and cost control.

Power electronics, electric motors, and control systems also require specialized manufacturing processes to ensure reliability under high-voltage and high-temperature conditions. Component standardization remains limited, which can result in longer production cycles and higher initial costs. Furthermore, the environmental impact of sourcing battery materials, including mining and processing, raises sustainability concerns that must be addressed throughout the lifecycle.

Operational and Performance Limitations

Electric aviation faces operational challenges, primarily due to energy storage limitations. Current battery technology provides far lower energy density compared to jet fuel, limiting aircraft range and endurance. While conventional commercial aircraft can fly thousands of kilometers without refueling, electric aircraft are typically constrained to short-haul or regional operations. High-power demand during takeoff and climb places additional stress on batteries, necessitating robust thermal management systems to prevent overheating. Charging infrastructure is another operational constraint, as high-capacity charging stations are not yet widely available at airports. This limitation affects turnaround times, flight scheduling, and scalability of electric air transport networks. Performance reliability is closely tied to battery health. Over time, batteries degrade, reducing energy capacity and efficiency. Unlike conventional engines, which can operate reliably for decades with maintenance, battery lifespan is finite, creating additional operational planning challenges for fleet operators.

Maintenance and Lifecycle Costs

Maintenance practices for electric aircraft differ substantially from conventional aircraft. Electric motors have fewer moving parts, reducing mechanical wear and lowering maintenance frequency. However, maintaining battery systems, thermal management components, and high-voltage electronics introduces new complexities. Battery inspection, monitoring, and replacement are critical to prevent in-flight failures and ensure safety.

The high cost of batteries and power electronics also contributes to lifecycle expenses. While electric propulsion may reduce fuel costs and mechanical servicing, the replacement and recycling of batteries can be expensive. Airlines and operators must plan for recurring capital expenditures to maintain fleet availability, potentially offsetting some of the operational savings from reduced fuel and maintenance requirements.

Environmental and Sustainability Considerations

Electric aviation is often promoted as a sustainable alternative due to its potential for zero in-flight emissions. However, the lifecycle environmental impact includes material extraction, battery production, energy generation, and end-of-life disposal. Mining for lithium, cobalt, and nickel can have significant environmental and social impacts. Battery manufacturing is energy-intensive, and electricity sources used for charging determine the net emissions. Recycling and disposal of batteries pose additional challenges. Current recycling technologies are limited, and improper disposal can lead to environmental contamination. Developing closed-loop recycling processes and using renewable energy for battery production and charging are essential to maximize the sustainability benefits of electric aviation.

Infrastructure and Regulatory Challenges

Electric aircraft require a robust infrastructure for charging, maintenance, and safety compliance. Airports need high-capacity charging systems, emergency response protocols for battery fires, and facilities for battery storage and replacement. Integrating these requirements into existing airport operations is costly and logistically complex. Regulatory frameworks for electric aviation are still evolving. Certification of electric propulsion systems, battery safety standards, and operational limits must be established to ensure safety and reliability. Unlike conventional aircraft, which have decades of established standards, electric aviation must navigate regulatory uncertainty, potentially slowing market adoption.

Technological Risks and Reliability

Electric aviation relies on emerging technologies that are not yet fully mature. Battery performance, thermal management, and power electronics can all experience failures under extreme conditions. Thermal runaway in batteries, electronic controller malfunctions, or motor overheating could have catastrophic consequences if not properly mitigated.

Redundancy and fault-tolerant design are crucial, but adding backup systems increases weight and complexity, partially offsetting the efficiency advantages of electric propulsion. Additionally, electric aircraft require sophisticated monitoring systems to detect early signs of failure, which can introduce new cybersecurity and data reliability concerns.

Economic and Market Factors

The economic viability of electric aviation depends on both initial acquisition costs and operational savings. While electric propulsion reduces fuel consumption and mechanical maintenance, the high cost of batteries, power electronics, and specialized infrastructure can create a substantial upfront investment. Market adoption is influenced by aircraft size, route length, and passenger demand.

Electric aircraft are currently most suitable for short-haul, regional, or urban air mobility markets. Scaling electric aviation to long-haul commercial flights will require breakthroughs in battery energy density, hybrid systems, or alternative fuels such as hydrogen. Economic incentives, government subsidies, and environmental regulations will play key roles in accelerating adoption.

Table: Lifecycle Challenges of Electric Aviation

Lifecycle Stage	Challenges
Design	Weight-energy trade-offs, stability, safety integration
Manufacturing	Material scarcity, complex assembly, component reliability
Operation	Limited range, battery degradation, charging infrastructure
Maintenance	Battery inspection, electronics servicing, high replacement cost
Environment	Mining impact, energy source emissions, recycling limitations
Regulation	Certification standards, safety protocols, operational guidelines
Infrastructure	Charging facilities, emergency response, airport adaptation
Technology	Thermal management, motor reliability, fault tolerance
Economics	High initial cost, energy cost, market viability
Market	Route length limits, fleet integration, passenger acceptance

Lifecycle Challenges of Electric Aviation

- **Battery Weight and Energy Density:** Current batteries are heavy relative to their energy content.
- **Thermal Management:** Preventing overheating of batteries and motors is critical.
- **Battery Degradation:** Energy capacity decreases with repeated charging cycles.
- **High Initial Cost:** Battery systems and power electronics are expensive.
- **Limited Infrastructure:** Charging stations and maintenance facilities are scarce.
- **Material Sourcing:** Mining of lithium, cobalt, and nickel has environmental and social impacts.
- **Recycling and Disposal:** End-of-life battery management is complex and costly.
- **Operational Range Limits:** Electric aircraft are generally constrained to short-haul missions.
- **Regulatory Uncertainty:** Certification and safety standards for electric aviation are still developing.
- **Technological Maturity:** Emerging technologies may face reliability and fault-tolerance challenges.

These challenges highlight that while electric aviation offers substantial benefits, its successful integration requires addressing multiple dimensions across the aircraft lifecycle.

Strategies for Mitigating Lifecycle Challenges

To overcome lifecycle challenges, manufacturers and operators are exploring several strategies. Advanced battery technologies, such as solid-state batteries, promise higher energy density and improved safety. Hybrid-electric configurations can extend range and reduce dependency on battery capacity while still lowering emissions.

Developing recycling and closed-loop supply chains for batteries can reduce environmental impact and lower long-term costs. Airports can invest in charging infrastructure, thermal management systems, and emergency response protocols to support electric operations. Collaborative efforts between manufacturers, regulators, and operators are essential to standardize certification and operational guidelines. Moreover, lifecycle assessment tools are being used to evaluate environmental and economic impacts, enabling informed decisions regarding aircraft design, materials, and energy sources. By addressing each stage of the lifecycle, electric aviation can achieve a sustainable, safe, and economically viable future.

2.3 POWER ELECTRONICS AND CONTROL SYSTEMS

Power electronics and control systems collectively form one of the most critical technological domains underpinning modern society's electrical infrastructure, industrial capabilities, transportation systems, and consumer products. The seamless integration of efficient power conversion with intelligent control enables technologies that are simultaneously more capable, more efficient, and more responsive to human needs and environmental constraints. As global trends toward electrification, digitalization, and sustainability intensify, the importance of advanced power electronics and sophisticated control strategies will only increase.

The ongoing development of new semiconductor materials, innovative circuit topologies, advanced control algorithms, and integrated system architectures promises to deliver continued improvements in performance, efficiency, and functionality across countless applications. The interdisciplinary nature of the field ensures that advances in related areas like materials science, computer engineering, and artificial intelligence will continue to catalyze innovation. Ultimately, power electronics and control systems represent enabling technologies that will play a central role in addressing some of society's most pressing challenges, including climate change, resource efficiency, and sustainable economic development, making them not merely technical disciplines but essential contributors to building a better technological future.

2.3.1 Fundamental Principles of Power Electronics

At its essence, power electronics involves the processing of electrical energy from one form to another using power semiconductor devices as switches. Unlike analog systems that dissipate significant power as heat, power electronic converters operate switches in either fully on or fully off states, minimizing losses and achieving high efficiency.

The foundational building blocks of power electronics include various converter topologies that perform specific conversion functions. Rectifiers convert alternating current to direct current, essential for power supplies and motor drives. Inverters perform the reverse operation, converting DC to AC, which is crucial for solar power systems, uninterruptible power supplies, and electric vehicle drives. DC-DC converters transform DC voltage levels, enabling power management in portable electronics, computers, and renewable energy systems. AC-AC converters directly change AC voltage magnitude or frequency, valuable for motor speed control and lighting applications. The operation of these converters relies on precise switching control of power semiconductor devices such as MOSFETs, IGBTs, and increasingly, wide bandgap semiconductors like silicon carbide and gallium nitride. These devices can handle substantial voltages and currents while switching at high frequencies, enabling compact, efficient power conversion systems.

2.3.2 Control Systems Theory and Applications

Control systems provide the theoretical framework and practical methodologies for managing the behavior of dynamic systems to achieve desired performance characteristics. These systems continuously measure output variables, compare them with reference inputs, and compute corrective actions to minimize error.

The field encompasses both classical control theory, which primarily deals with single-input single-output systems using frequency domain techniques, and modern control theory, which addresses multiple-input multiple-output systems using time-domain state-space approaches. Control systems are broadly categorized into open-loop systems, which operate without feedback, and closed-loop systems, which utilize feedback to maintain performance despite disturbances and uncertainties.

Proportional-Integral-Derivative controllers represent one of the most widely implemented control strategies due to their effectiveness and relative simplicity. More advanced techniques include adaptive control, which adjusts controller parameters in response to changing system dynamics; robust control, designed to maintain performance despite model uncertainties; and optimal control, which minimizes a performance index. Modern applications increasingly employ intelligent control methods using artificial intelligence, fuzzy logic, and neural networks to handle complex, nonlinear systems where conventional approaches may be inadequate.

2.3.3 Components and Technologies

The implementation of effective power electronics and control systems relies on several critical components and technologies that collectively enable precise power management and system control.

- **Power Semiconductor Devices:** These form the heart of power electronic systems, with IGBTs, MOSFETs, and emerging wide bandgap semiconductors like SiC and GaN providing the switching capability essential for efficient power conversion.
- **Microcontrollers and Digital Signal Processors:** These integrated circuits serve as the computational backbone of modern control systems, executing complex algorithms in real-time to regulate system behavior.

- **Sensors and Measurement Systems:** Accurate current, voltage, temperature, speed, and position sensors provide the essential feedback signals that enable closed-loop control and system monitoring.
- **Magnetic Components:** Inductors and transformers provide energy storage, filtering, and voltage transformation capabilities critical for power conversion circuits.
- **Capacitors and Energy Storage Elements:** These components filter switching ripple, provide temporary energy storage, and stabilize voltage levels within power electronic systems.
- **Gate Drive Circuits:** Specialized circuits translate control signals from processors to the voltage and current levels required to efficiently switch power semiconductor devices.
- **Communication Interfaces:** Standardized protocols like CAN, Ethernet, and wireless links enable coordination between distributed power electronic systems and higher-level control networks.
- **Heat Management Systems:** Heat sinks, thermal interface materials, and active cooling systems dissipate waste heat generated by power losses in semiconductor devices.
- **Protection Circuits:** Overcurrent, overvoltage, overtemperature, and short-circuit protection mechanisms ensure system reliability and prevent catastrophic failures.
- **Software and Programming Tools:** Development environments, compilers, and specialized software enable the implementation, testing, and tuning of control algorithms for specific applications.

2.3.4 Applications in Renewable Energy Systems

Power electronics and control systems play an indispensable role in harnessing, managing, and integrating renewable energy sources into modern power systems. In photovoltaic solar installations, power electronic converters perform maximum power point tracking to extract the highest possible energy from solar panels despite varying irradiation conditions, while also converting DC power to grid-compatible AC power.

Modern solar inverters incorporate sophisticated control algorithms that not only manage power conversion but also provide grid support functions like voltage regulation, frequency stabilization, and fault ride-through capability. Wind energy systems similarly rely on advanced power electronics to interface variable-speed turbines with the electrical grid, enabling efficient operation across a wide range of wind speeds. The control systems in wind turbines manage blade pitch, generator torque, and power flow to maximize energy capture while minimizing mechanical stresses. At the system level, power electronics enable the integration of energy storage systems with renewable generation, facilitating time-shifting of energy availability and enhancing grid stability. Grid-forming inverters represent a particularly advanced application, where power electronic systems with sophisticated controls can actually establish and maintain grid voltage and frequency, enabling the creation of microgrids and enhancing the resilience of distributed energy systems.

2.3.5 Industrial Applications and Motor Drives

Industrial applications represent one of the most significant domains for power electronics and control systems, particularly in the area of electric motor drives. Adjustable speed drives using power electronic converters control the voltage and frequency supplied to AC motors, enabling precise speed regulation and substantial energy savings compared to traditional fixed-speed operation.

These systems incorporate sophisticated control algorithms like field-oriented control and direct torque control that provide performance characteristics rivaling those of DC motors while maintaining the robustness and low maintenance of AC induction or synchronous motors.

Beyond motor control, power electronics provide regulated power supplies for industrial processes, uninterruptible power systems for critical loads, and high-power rectifiers for electrochemical applications. Control systems manage complex industrial processes ranging from chemical production and material handling to manufacturing automation and robotics. Programmable logic controllers combined with power electronic actuators enable the automation of industrial processes with high precision, repeatability, and efficiency. The integration of power electronics with industrial control networks and the Industrial Internet of Things is creating smarter, more connected factories where energy usage is optimized, maintenance is predictive, and production flexibility is enhanced.

2.3.6 Transportation Electrification Systems

The ongoing electrification of transportation systems represents one of the most visible and impactful applications of advanced power electronics and control technologies. In electric and hybrid vehicles, power electronic converters manage energy flow between the battery, motor, and auxiliary systems. The traction inverter, a sophisticated power electronic system, converts DC battery power to controlled AC power for the propulsion motor, with advanced control algorithms regulating torque and speed with exceptional precision. DC-DC converters step voltage levels to power different vehicle systems, while onboard chargers manage the interface with external power sources. Control systems in electric vehicles perform battery management, thermal management, regenerative braking control, and vehicle dynamics optimization. Beyond road vehicles, power electronics and control systems are equally critical in electric trains, where they manage power collection from overhead lines or third rails and control traction motors. The emerging field of electric aviation similarly relies on specialized power electronic systems to manage propulsion, power distribution, and thermal management in aircraft. In all transportation applications, the control systems must ensure safety, reliability, and efficiency under demanding operational conditions, while power electronics must deliver high power density and efficiency within strict weight and volume constraints.

2.3.7 Smart Grid and Power Quality Applications

The modernization of electrical grids into smart, responsive networks is fundamentally enabled by advances in power electronics and control systems. Flexible AC transmission systems use power electronic controllers to enhance the controllability, stability, and power transfer capability of existing transmission networks.

Distribution static compensators and dynamic voltage restorers maintain power quality by mitigating voltage sags, swells, and harmonic distortions at the distribution level. Active power filters employ power electronic converters with sophisticated control algorithms to cancel harmonic currents and reactive power, improving overall power factor and reducing losses. The grid interface of renewable energy sources, energy storage systems, and electric vehicle charging infrastructure all rely on power electronic converters with grid-supporting controls that can respond to grid conditions and operator commands.

Advanced metering infrastructure incorporates power electronics for energy measurement and communication interfaces, while control systems manage the data acquisition, processing, and communication functions. At the consumer level, power electronics enable efficient power supplies for electronic devices, lighting controls, and home energy management systems. The coordination of these diverse applications through hierarchical control architectures is transforming passive electrical grids into active, intelligent networks that optimize energy delivery, enhance reliability, and accommodate diverse generation and load patterns.

2.3.8 Challenges and Research Frontiers

Despite significant advances, numerous challenges and research frontiers remain in power electronics and control systems. Reliability and lifetime prediction continue to be complex issues, particularly for new semiconductor technologies and applications with demanding operational profiles. The design of control systems that can maintain stability and performance despite component aging, parameter variations, and external disturbances represents an ongoing research challenge. Thermal management remains a limiting factor for power density, especially in high-power applications where heat flux densities continue to increase.

Electromagnetic compatibility is increasingly difficult to maintain as switching speeds rise and system densities increase, requiring sophisticated modeling and mitigation techniques. The development of accurate models that capture the multi-physics nature of power electronic systems—combining electrical, thermal, and mechanical domains—remains computationally challenging. For control systems, managing the complexity of large-scale, networked systems while ensuring stability and performance presents theoretical and practical difficulties.

The cybersecurity of networked power electronic systems is an emerging concern, particularly for critical infrastructure applications. Cost reduction while maintaining performance is a perpetual challenge, especially for emerging technologies like wide bandgap semiconductors. Interdisciplinary approaches that combine expertise from materials science, electrical engineering, computer science, and control theory are increasingly necessary to address these complex challenges and advance the state of the art.

2.4 AIRCRAFT ENERGY MANAGEMENT SYSTEMS (AEMS)

The next wave of AEMS innovation blends better chemistry-agnostic models with tighter integration across the aircraft. Solid-state batteries and fuel cells will arrive incrementally, but AEMS will deliver much of the near-term benefit by exploiting existing hardware more intelligently. Expect standardized high-voltage architectures, shared thermal loops with active heat pumps, and fleet-level optimization that assigns aircraft to routes based on real-time SOH and weather. As regulators codify energy reserve rules and verification methods, data-rich AEMS will unlock smaller buffers without compromising safety. In short, AEMS is the control system that makes electrified aviation practical: turning electrons, heat, and time into predictable missions, healthy components, and sustainable economics.

2.4.1 AEMS is and Why it Matters

At its core, AEMS integrates multiple energy sources (generators, batteries, fuel cells, supercapacitors), buses (AC and DC at various voltages), and loads (propulsors, environmental control systems, flight controls, avionics, de-icing, lighting) through a supervisory controller. The system monitors energy state, predicts future demand along the planned trajectory, and makes allocation decisions under safety, thermal, and regulatory constraints. In conventional aircraft, fuel served as a flexible energy buffer with high specific energy and simple metering; electrified aircraft move complexity into power electronics and software. That shift demands precise estimation of state-of-charge (SOC), state-of-energy (SOE), and state-of-health (SOH), plus intelligent strategies for power split, load shedding, and reserve protection. AEMS thereby becomes the key lever to extend range, increase dispatch reliability, reduce lifecycle cost, and meet certification goals.

2.4.2 Architecture: Sources, Buses, and Loads

AEMS sits atop a power architecture often called electrical power generation and distribution (EPGD) or power management and distribution (PMAD). On hybrid platforms, turbogenerators feed high-voltage DC buses (typically 540–1000 VDC), with inverters supplying propulsors and DC/DC converters stepping down to 28 VDC for legacy systems. More-electric aircraft may retain 115 VAC, 400 Hz or variable frequency AC buses alongside HVDC to support standardized equipment. Batteries connect through contactors and fault protection to the HVDC bus, sometimes segmented into zones with bus ties for redundancy. Supercapacitors or high-power buffer modules can shave transient loads like eVTOL liftoff. Downstream, point-of-load converters ensure quality power (voltage, frequency, total harmonic distortion) to sensitive avionics.

Load classification under AEMS is explicit: safety-critical (flight controls, avionics), mission-critical (propulsion, environmental control), and non-critical (galley, cabin amenities). This classification drives load-shedding ladders and minimum functionality after failures. Protection is layered: arc-fault detection, current limiting, zonal isolation, and lightning/EMI filtering per DO-160. The architecture must tolerate single-point failures—such as a generator dropout or battery isolation—without losing controllability, which pushes redundancy into both hardware paths and control logic.

2.4.3 Sensing, Estimation, and Digital Twins

Accurate energy awareness is the foundation of AEMS. Battery SOC alone is insufficient; SOE captures the interplay between voltage sag, temperature, and usable capacity under expected power demand. Estimators fuse current integration (coulomb counting), open-circuit voltage models, and impedance spectroscopy, often using Kalman or particle filters to manage noise and drift. SOH tracks degradation (loss of lithium inventory, impedance rise) to predict remaining useful life and adjust mission limits.

For generators and fuel cells, AEMS models spool dynamics, fuel flow, and efficiency maps to schedule setpoints that minimize fuel burn and thermal stress. Sensors measure bus voltage, current, temperature at cell/module/pack granularity, contactor states, and vibration on rotating machines. A digital twin—an onboard physics-based model calibrated with fleet data—forecasts energy demand across the flight plan given weather, winds aloft, traffic constraints, and predicted hover/

climb durations. This twin quantifies uncertainty bands, allowing AEMS to enforce reserves with a confidence level rather than a blunt fixed margin, reclaiming usable energy when conditions are benign.

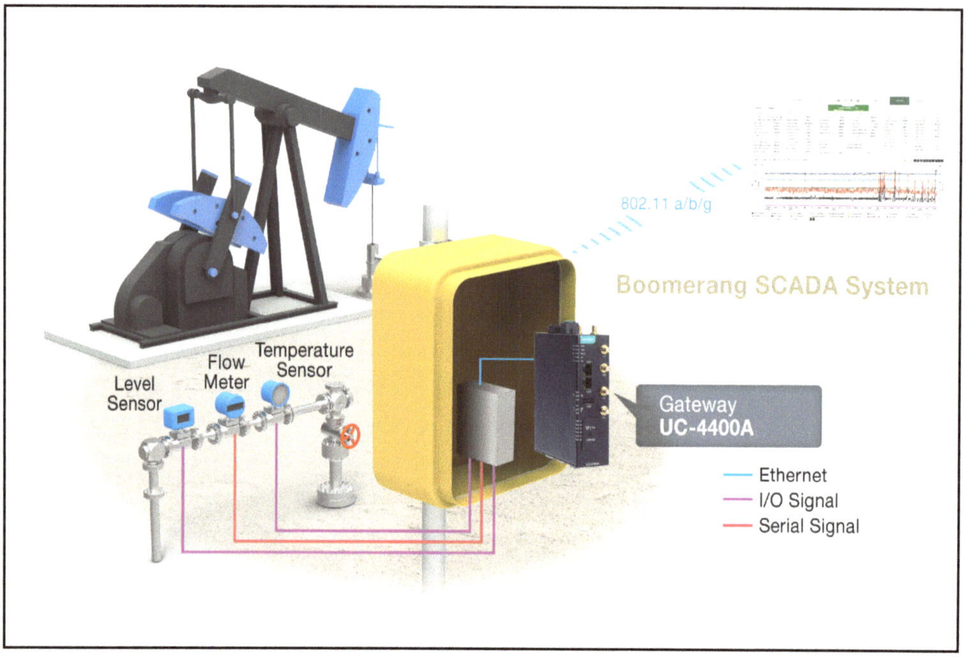

Fig: Sensing Systems, Data Estimation, and Virtual Twins

2.4.4 Control and Optimization Strategies

AEMS employs layered control. Fast loops in power electronics regulate currents and voltages; mid-level controllers manage component operating points; and a supervisory optimizer coordinates sources and loads. For hybrids, power split control blends generator and battery power using model predictive control (MPC) or rule-based strategies, respecting generator best-efficiency islands, battery C-rate limits, and thermal constraints.

During high-power phases (takeoff, go-around, eVTOL hover), batteries may provide the peak while generators handle baseload; in cruise, generators recharge within thermal limits if margins permit. Load management enforces priorities and schedules flexible consumers (cabin heat pumps, galley, icing protection) to avoid coincident peaks that stress the bus. Fault accommodation reconfigures buses and reassigns power in milliseconds, preserving controllability while avoiding cascading trips. Critically, AEMS interfaces with the flight management system (FMS) to enable energy-aware routing: adjusting climb profiles, speed, and altitude to land with target reserves and optimal thermal states. The objective is not simply minimizing energy; it is maximizing mission reliability and asset life under constraints.

2.4.5 Thermal Management is Energy Management

Heat is the tax on every watt. Batteries, inverters, motors, generators, and power electronics all produce waste heat that must be moved and rejected. AEMS coordinates thermal loops—liquid jackets, cold plates, oil spray, and heat exchangers—with ambient conditions and mission phases. Preconditioning batteries before departure restores power capability in cold weather and curbs aging during fast charge.

During hover or climb, AEMS may cap power to stay within thermal envelopes, trading time-to-climb for thermal headroom. Integrating thermal storage—phase-change materials or buffer volumes—can ride through short peaks without oversizing radiators, especially important for eVTOLs operating from hot-and-high vertiports. Cabin conditioning competes for thermal capacity; AEMS schedules compressor loads to off-peak times or modulates them with propulsor power to smooth total draw. Because thermal limits often derate electrical components before energy runs out, the tight coupling of thermal and electrical models is central to extracting maximum usable energy.

2.4.6 Avionics Integration and Human–Machine Interface

Pilots need a clear, predictive picture of energy, not just a percentage gauge. AEMS integrates with the FMS, navigation, and autopilot to present "energy to destination," alternate options, and green bands for speed/altitude that keep reserves intact. For eVTOLs, hover-time margins and diversion paths are visualized as geo-fenced envelopes. Alerts are progressive: advisory when trending toward reserve breach, caution when action is required (e.g., reduce speed, lower climb rate), and warning when minimum energy state is threatened.

The human–machine interface (HMI) must balance automation with transparency. Pilots should understand why AEMS recommends a power split or a speed change and retain authority to override with clear consequences displayed. For operations, AEMS shares post-flight energy and thermal data to inform turnaround plans, charge schedules, and maintenance. In training, simulators replicate AEMS logic so crews internalize energy-managed procedures just as they learn engine-out profiles today.

2.4.7 Safety, Reliability, and Certification

Safety is engineered into AEMS through standards and architecture. Development follows ARP4754A for system engineering and ARP4761 for safety assessment, with software qualified to DO-178C at appropriate Design Assurance Levels (DAL), hardware to DO-254, and environmental tests per DO-160. Battery systems adhere to DO-311A for rechargeable lithium batteries, demonstrating containment, venting, and non-propagation. Integrated modular avionics interfaces may follow DO-297. For eVTOL and hybrid aircraft, EASA SC-VTOL and FAA Part 23/27/29 guidance define energy reserve policies and reliability expectations. Architecturally, AEMS avoids single points of failure via redundant sensors, dissimilar processors for voting, independent power paths, and fail-operational strategies for critical functions. Fault detection and isolation (FDI) must be fast and selective—arc-fault and ground-fault protection that trips the smallest necessary segment. Cybersecurity is now mandatory; DO-326A/ED-202A and DO-355 guide threat assessment and aircraft network security so that energy controls cannot be spoofed or disrupted.

2.4.8 Ground Infrastructure and Lifecycle Economics

AEMS extends to the ramp. It negotiates with chargers over standardized protocols, manages precondition and charge profiles to minimize degradation, and arbitrages tariffs when on-site storage or PPAs are available. Turn time matters: fast charging raises cell temperature and ages packs; AEMS balances charge rate against remaining schedule and SOH impact, quantifying dollars per minute saved.

For operators, battery-as-a-service and separate asset accounting depend on AEMS telemetry—cycle counts, depth-of-discharge histograms, temperature exposure, and projected remaining useful life—to optimize swap timing and residual value. Economically, better AEMS means fewer energy-related delays, higher aircraft utilization, and longer component life. By flattening power peaks and coordinating thermal loads, it allows smaller, lighter components to pass certification without conservative oversizing. Accurate energy prediction lets dispatch choose routes and alternates that meet reserves with less buffer, effectively increasing payload or range without changing hardware.

2.4.9 Cases Across Aircraft Classes

- **eVTOLs and Urban Air Mobility:** Highly variable, peak-heavy profiles make AEMS essential. It minimizes hover time, protects thermal limits in hot-and-high environments, and expresses energy margins as time-in-hover equivalents that pilots intuitively manage.
- **Regional Hybrid-Electric:** AEMS schedules generator setpoints to operate near best-efficiency islands, uses the battery for climb and contingencies, and opportunistically recharges in descent while staying within noise and thermal constraints at small airports.
- **More-Electric Turbofans:** Even without battery propulsion, AEMS coordinates high-voltage DC distribution, replacing pneumatic and hydraulic loads with electric actuators and pumps. Load shedding and bus reconfiguration keep critical systems powered after a generator fault.

- **UAVs and HALE Platforms:** Energy-aware routing driven by winds aloft and solar input (for solar-electrics) extends endurance; AEMS balances battery SOC with solar harvest and payload demands, ensuring safe nighttime reserves.

2.4.10 Ten Practical AEMS Levers to Improve Range, Reliability, and Life

- Use model predictive control for power split in hybrids to keep generators in their efficiency sweet spots and cap battery C-rate during hot phases.
- Precondition batteries to a temperature window that maximizes both power and longevity; automate this with charger–AEMS handshakes on the ramp.
- Schedule flexible loads (cabin compressors, de-icing, galley) to avoid coincident peaks; a few seconds of offset can eliminate an inverter or cable upsizing.
- Raise bus voltage within certification limits to cut current, shrinking harness mass and I^2R losses while improving voltage stability under transients.
- **Instrument Deeply:** Cell-level voltage/temperature sensing and impedance tracking enable tighter SOE estimates and smaller operational reserves.
- Design explicit load-shedding ladders tied to mission phase so that non-critical loads drop automatically before energy margins erode.
- Integrate thermal storage or oversized cold plates where peak duration is short; this defers heavy, draggy radiators and preserves hover capability.
- Feed AEMS forecasts into the FMS so climb, speed, and altitude are energy-optimal, not just time-optimal—especially in headwinds.
- **Quantify Battery Depreciation Per Mission:** let dispatch trade a faster turnaround (higher charge rate) against marginal life cost with data.
- **Test for Cybersecurity Early:** Protect charger links, maintenance ports, and networked sensors so energy commands cannot be altered or delayed.

2.5 EMERGING TECHNOLOGIES

2.5.1 Supercapacitors

While they may never replace the high-energy battery packs required for sustained flight, supercapacitors are poised to become an essential "enabling technology" for practical electric aviation. They solve the difficult problems of thermal management, transient power response, and system longevity that currently plague pure-battery architectures. As eVTOLs and larger electric fixed-wing aircraft move from prototypes to commercial service, the hybridized approach—using batteries for the marathon and supercapacitors for the sprints—will likely become the standard specifically to ensure the reliability, safety, and economic viability required for certified commercial operations.

Hybrid Architecture Advantage

The most promising application for supercapacitors in electric aviation is within a hybridized energy storage system. A "pure" supercapacitor aircraft would have spectacular takeoff performance but might only fly for a few minutes due to their low total energy storage. Conversely, a pure battery aircraft might have decent range but would require oversized, heavy battery packs just to survive the takeoff phase without overheating.

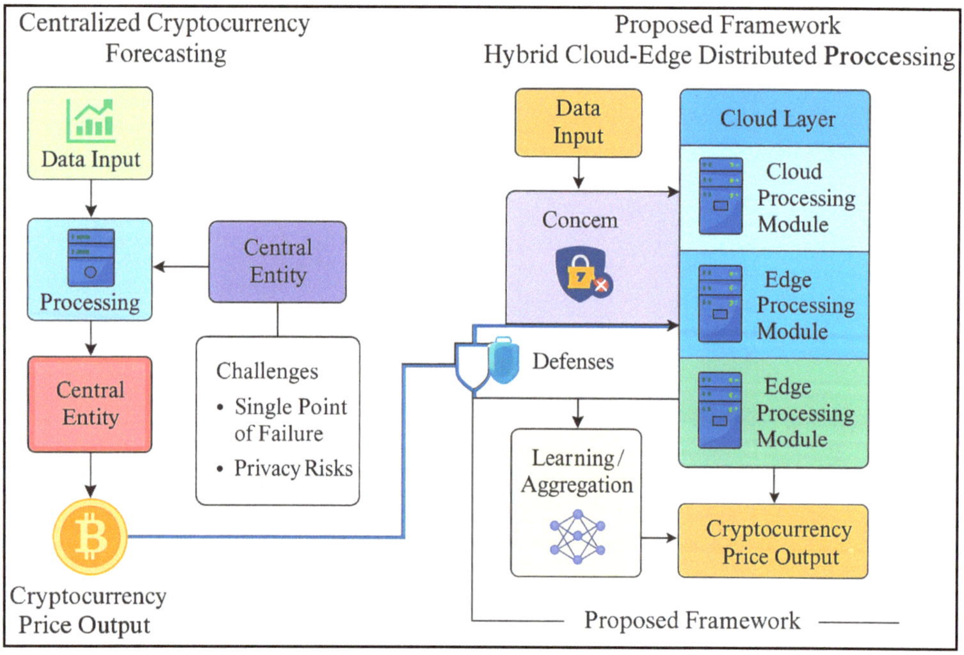

Hybrid Cloud–Edge Architecture for Real-Time Cryptocurrency Market Forecasting

By combining the two, engineers can "right-size" both components. The battery can be optimized purely for cruise energy density (endurance), while a relatively small, lightweight supercapacitor pack handles the peak power demands. This synergy can actually lead to a lighter overall energy system with a longer operational life, as the expensive main battery is buffered from the most damaging phases of flight.

Physics of Power versus Energy

To understand the role of supercapacitors in aviation, one must distinguish between energy density and power density. If a flight is a marathon, the battery is the runner with immense stamina, capable of jogging for hours. The supercapacitor is the sprinter, capable of explosive speed but only for a few seconds. In standard electric aircraft architectures, asking the main battery to perform maximum-power takeoffs repeatedly is akin to asking a marathon runner to sprint a 100-meter dash every few miles; it causes rapid fatigue and injury. Supercapacitors store energy in an electric field between two conductive plates separated by an insulating electrolyte, rather than in chemical bonds. Because no chemical phase change occurs during discharge, they can cycle millions of times with virtually no degradation, regardless of how quickly they are discharged. In an aircraft, this means they can provide the massive surge of current needed to spin up propellers from a standstill to takeoff RPM in seconds, protecting the main battery from the high-current stress that typically causes lithium plating, overheating, and capacity fade.

Specific Roles and Utility in Flight Operations

The integration of supercapacitors solves several nuanced engineering challenges that are unique to the electric aviation environment. Their ability to react instantly to power demands makes them indispensable for modern, highly responsive aircraft designs. Their specific utility in next-generation aircraft can be broken down into several critical functional areas that enhance performance, safety, and economics. The following ten points highlight exactly where supercapacitors fit into the electric aviation ecosystem:

- **Peak Power Shaving for Takeoff:** Supercapacitors can supply the massive, short-duration current spike required during the initial takeoff roll and climb-out, significantly reducing the thermal load and C-rate stress placed on the main lithium-ion battery pack during its most vulnerable operational phase.
- **Regenerative Braking Capture:** During rapid descent or deceleration, propellers can act as wind turbines, generating large surges of electrical energy. Batteries often cannot absorb this “regen” energy quickly enough without damage, whereas supercapacitors can capture nearly 100% of this transient energy for later use.
- **Hydrogen Fuel Cell Load Leveling:** Hydrogen fuel cells prefer to operate at a steady, constant output and are very slow to react to sudden throttle changes. Supercapacitors act as a necessary buffer, instantly supplying power when the pilot advances the throttle while waiting for the fuel cell to “spool up” to the new demand level.

- **eVTOL Hover Stabilization:** Maintaining a steady hover in turbulent air requires motors to make hundreds of micro-adjustments per second, demanding erratic, rapid-fire bursts of power. Supercapacitors are perfectly suited to handle this high-frequency "noise" in power demand that would otherwise strain a battery.
- **Low-Temperature Performance Assurance:** Lithium-ion batteries suffer severe performance drops in cold weather, losing much of their ability to provide power. Supercapacitors are largely unaffected by cold down to roughly -40°C, ensuring reliable aircraft startup and initial power availability even in freezing conditions.
- **Emergency Actuation Power:** In the event of a total main battery failure, supercapacitors can serve as an ultra-reliable, instantly available emergency power source to drive critical flight control surfaces, deploy landing gear, or power essential avionics for a safe emergency landing.
- **Main Battery Cycle Life Extension:** By handling all the high-stress transient loads, supercapacitors allow the main battery to operate in a gentle, steady-state discharge mode. This can potentially double or triple the usable cycle life of expensive flight battery packs, drastically improving aircraft operating economics.
- **Voltage Sag Mitigation:** When high-power electric motors draw huge currents, the entire aircraft's system voltage can momentarily drop (sag), potentially causing sensitive avionics to reboot. Supercapacitors act as a stiff voltage buffer, maintaining a stable electrical bus even during extreme motor loads.
- **Rapid Ground Charging Support:** While they can't store complete flight energy, on-board supercapacitors could be flash-charged in seconds while passengers board to provide just enough energy for taxiing and takeoff, leaving the main battery fully charged for the actual flight portion.
- **Safety Margin Enhancement:** Because their state-of-charge can be known with extremely high accuracy (unlike batteries, whose remaining charge is sometimes an estimate), supercapacitors provide a highly reliable "known quantity" of reserve energy that pilots can depend on for critical last-resort maneuvers.

Challenges and Material Limitations

Despite their immense benefits in power handling, supercapacitors have significant limitations that currently restrict their broader adoption. The primary drawback is their extraordinarily low specific energy—often less than 10 Wh/kg, compared to

250+ Wh/kg for modern aviation batteries. This means they are essentially "dead weight" during the cruise phase of flight when steady, long-duration power is needed. Adding them to an aircraft increases complexity, requiring specialized DC-DC converters to manage the energy flow between the supercapacitor, the battery, and the motors. Furthermore, supercapacitors have a high self-discharge rate; if an aircraft is left parked for several days, the supercapacitors might bleed off much of their stored energy, whereas the main batteries would remain stable.

2.5.2 Hydrogen Fuel Cells

Hydrogen fuel cells represent a transformative technology for electric aviation, offering a pathway to sustainable, low-emission flight. By converting hydrogen into electricity with water as the only byproduct, fuel cells provide a clean, quiet, and efficient alternative to conventional jet engines. However, the lifecycle of hydrogen fuel cell aircraft is complex, involving challenges in storage, thermal management, infrastructure, safety, regulation, economics, and material sourcing. Challenges requires coordinated efforts from manufacturers, regulators, infrastructure developers, and energy providers. Hybrid-electric configurations, advanced storage solutions, and green hydrogen production are key enablers of practical fuel cell aviation. As these technologies mature, hydrogen fuel cells could play a central role in the decarbonization of air travel, enabling a future of safe, efficient, and environmentally sustainable aviation. The successful integration of hydrogen fuel cells into aviation is not merely a technological milestone but a strategic shift toward meeting global climate goals and redefining the environmental footprint of flight. With ongoing research, investment, and regulatory development, hydrogen fuel cell electric aviation may soon become a viable and competitive alternative to conventional aircraft propulsion.

Hydrogen Fuel Cell Technology

Hydrogen fuel cells operate by converting the chemical energy of hydrogen directly into electricity without combustion. The primary components of a fuel cell include the anode, cathode, electrolyte, and associated bipolar plates. Hydrogen gas is supplied to the anode, where it splits into protons and electrons. The electrons travel through an external circuit, generating electrical current, while the protons pass through the electrolyte to the cathode, where they combine with oxygen to form water. This process is highly efficient and produces no harmful emissions. Fuel cells in aviation can be used to power electric motors directly or in combination with battery systems in hybrid-electric architectures. Unlike conventional jet fuel engines, hydrogen fuel cells offer a clean energy source and can potentially reduce noise, maintenance requirements, and operational costs. The technology is particularly suited for short-haul and regional aircraft, where energy demands are moderate, and environmental benefits are most impactful.

Advantages of Hydrogen Fuel Cells in Aviation

Hydrogen fuel cells offer several advantages over traditional propulsion systems and even conventional batteries.

- **Emissions Reduction is a Major Benefit:** Fuel cells produce only water vapor during operation, eliminating carbon dioxide, nitrogen oxides, and particulate matter associated with fossil fuel combustion. This makes hydrogen fuel cells a key enabler of zero-emission aviation.
- **Fuel Cells Provide High Energy Efficiency:** Converting chemical energy to electrical energy at rates of 50–60%, which can be higher when combined with energy recovery systems. Unlike batteries, hydrogen storage allows for greater energy density per kilogram, enabling longer flight ranges without the weight penalties of heavy batteries.
- **Another Advantage is Quiet Operation:** Electric motors powered by fuel cells generate significantly less noise than combustion engines, reducing noise pollution near airports and urban areas. Furthermore, fuel cells have fewer moving parts compared to traditional engines, resulting in lower maintenance requirements and potentially higher reliability over the lifecycle of the aircraft.

Hydrogen Storage and Supply Challenges

Despite the advantages, hydrogen fuel cells introduce unique challenges, particularly in terms of storage and supply. Hydrogen has a low volumetric energy density, which requires either high-pressure gas storage, cryogenic liquid storage, or chemical carriers. High-pressure tanks must withstand extreme pressures of up to 700 bar, while cryogenic storage involves cooling hydrogen to -253°C, creating insulation and safety challenges. The infrastructure for producing, transporting, and storing hydrogen is still in its infancy. Airports must develop refueling stations, safe handling protocols, and supply chains to ensure consistent availability of hydrogen. The source of hydrogen also affects the sustainability of the technology. Hydrogen produced from renewable energy through electrolysis is considered green hydrogen, while hydrogen derived from natural gas via steam reforming (grey hydrogen) still carries associated carbon emissions.

Integration of Hydrogen Fuel Cells in Aircraft

Integrating hydrogen fuel cells into aircraft requires innovative design strategies. Fuel cells can power electric motors directly or provide electricity to batteries that act as buffers for high-power demands during takeoff and climb. Hybrid-electric architectures help balance energy supply, weight, and performance, allowing fuel cells to operate efficiently while meeting peak power requirements. Distributed propulsion concepts, where multiple smaller motors are placed along the wings or fuselage, are particularly compatible with hydrogen fuel cells. This arrangement enhances aerodynamic efficiency, redundancy, and noise reduction. Aircraft such as Airbus ZEROe and regional hydrogen-powered prototypes are exploring these integration strategies, aiming to demonstrate the feasibility of hydrogen-electric propulsion in commercial aviation.

Table: Hydrogen Fuel Cells vs. Conventional Aviation

Parameter	Hydrogen Fuel Cells	Conventional Aviation
Energy Source	Hydrogen gas	Jet fuel or avgas
Efficiency	50–60% (higher with hybridization)	30–40% thermal efficiency
Emissions	Water vapor only	CO_2, NO_x, particulates
Noise	Low	High engine and exhaust noise
Maintenance	Fewer moving parts	Complex, frequent servicing
Weight Impact	Hydrogen storage systems add weight	Fuel weight decreases as consumed
Range	Moderate to long with hybrid systems	Long-haul capable
Refueling Infrastructure	Limited, developing	Established global network
Sustainability	High potential with green hydrogen	Environmentally unsustainable
Safety	Requires pressure and temperature management	Proven technology with fuel handling protocols

Performance Considerations

Hydrogen fuel cells enable high efficiency and smooth electric propulsion, but performance depends on the power-to-weight ratio and energy density of the storage system. High-pressure or cryogenic tanks add weight and occupy space, impacting payload capacity and cabin configuration. Optimizing the aircraft structure and integrating lightweight materials is crucial to maintaining aerodynamic performance.

Fuel cells perform efficiently at steady power output, making them well-suited for cruise phases of flight. However, peak power requirements during takeoff and climb may require supplementary energy sources such as batteries or hybrid systems. This ensures safe and reliable operations while preventing fuel cell degradation from rapid load changes. Temperature extremes at high altitudes also affect fuel cell performance. Thermal management systems are necessary to maintain optimal operating conditions, ensuring efficiency and reliability across the flight envelope.

Safety and Regulatory Challenges

Safety is paramount in aviation, and hydrogen introduces unique considerations. Hydrogen is highly flammable and requires careful handling, leak detection, and ventilation systems to prevent accidents. Aircraft must incorporate safety features such as pressure relief valves, crash-resistant storage, and automated shutdown systems. Regulatory standards for hydrogen fuel cell aircraft are still evolving. Certification requires rigorous testing of storage, propulsion, and operational systems to ensure safety in diverse flight conditions. Regulatory bodies such as the European Union Aviation Safety Agency (EASA) and Federal Aviation Administration (FAA) are developing guidelines to address hydrogen-specific risks, including high-pressure containment, thermal management, and emergency response procedures.

Economic Considerations

Economic viability is a critical factor for hydrogen fuel cell aviation. Initial aircraft acquisition costs are high due to the price of fuel cell systems, hydrogen storage tanks, and integration with electric motors. Operational costs can be lower than conventional aircraft if hydrogen production and distribution are efficient, and maintenance requirements are reduced due to fewer moving parts. Fuel cost stability is another advantage. Hydrogen prices are less susceptible to the volatility of fossil fuel markets, especially if green hydrogen production scales. Government incentives and environmental regulations may further enhance the economic attractiveness of hydrogen fuel cell aircraft by supporting early adoption in short-haul and regional markets.

Environmental and Sustainability Implications

Hydrogen fuel cells offer a pathway to near-zero-emission aviation, particularly when using green hydrogen produced from renewable electricity. This reduces carbon footprints, air pollution, and dependence on fossil fuels. Additionally, the quiet operation of electric motors powered by fuel cells minimizes noise pollution in urban and regional airports. Sustainability depends on the entire lifecycle, including hydrogen production, storage, and end-of-life management of fuel cell systems. Recycling of fuel cell materials, such as platinum catalysts and membranes, is essential to reduce environmental impact and maintain resource efficiency. Proper management ensures that hydrogen fuel cell aviation contributes meaningfully to the decarbonization of the aviation sector.

Hydrogen Fuel Cell Aviation

- **Hydrogen Storage:** Requires high-pressure tanks or cryogenic systems, adding weight.
- **Energy Density:** Lower volumetric energy density compared to jet fuel.
- **Thermal Management:** Maintaining optimal operating temperature for fuel cells.

- **Infrastructure:** Limited refueling networks at airports.
- **Safety Risks:** Flammability of hydrogen requires robust safety systems.
- **Regulatory Standards:** Certification frameworks are still under development.
- **Material Availability:** Fuel cell components rely on critical materials like platinum.
- **Cost:** High initial investment in aircraft, storage, and refueling systems.
- **Hybrid Integration:** Necessary for peak power demands during takeoff and climb.
- **Lifecycle Management:** Recycling and end-of-life handling of fuel cells is complex.

These challenges indicate that while hydrogen fuel cells hold significant promise, achieving commercial viability requires overcoming technological, regulatory, and infrastructure barriers.

Applications and Prospects

Hydrogen fuel cells are particularly suitable for regional and short-haul flights, where energy demands are moderate, and emissions reduction has a high impact. Urban air mobility vehicles, commuter aircraft, and training planes can benefit from fuel cell systems due to low noise, reliability, and reduced operational costs. Longer-range applications will require hybrid architectures or innovations in hydrogen storage, such as metal hydrides or ammonia carriers, which increase energy density. Manufacturers like Airbus, ZeroAvia, and Universal Hydrogen are actively developing prototypes and demonstration aircraft to test fuel cell performance, storage solutions, and refueling logistics.

As technology matures, hydrogen fuel cell aircraft could expand into regional transport networks and potentially medium-haul flights, complementing battery-electric and conventional propulsion systems. The ultimate vision is a fully sustainable aviation sector with multiple zero-emission propulsion technologies integrated according to route, payload, and operational requirements.

2.5.3 Wireless Charging

Wireless charging technology represents far more than a simple convenience for electric aviation; it is a pivotal enabling technology with the potential to fundamentally reshape aircraft operations and infrastructure economics. By providing a safe, automated, and efficient method for powering electric aircraft, wireless charging directly addresses one of the most critical challenges in the sector: maintaining high aircraft utilization with minimal ground time. The technology offers compelling advantages across multiple dimensions, including operational efficiency, safety enhancement, automation compatibility, and environmental sustainability. While significant challenges remain in terms of cost, technical performance, and regulatory framework development, the rapid pace of innovation and growing industry commitment suggest these hurdles will be progressively addressed in the coming years. Current implementation projects and demonstration programs are providing valuable validation of the technology's readiness and benefits, building confidence among operators, manufacturers, and investors. As electric aviation continues its transition from concept to commercial reality, wireless charging infrastructure will play an increasingly strategic role in determining the operational and economic viability of new aircraft types and business models. The successful development and deployment of wireless charging systems will therefore be essential for realizing the full potential of electric aviation to deliver sustainable, efficient, and accessible air transportation for the future.

Fundamental Technologies and Operating Principles

Wireless charging for aviation primarily leverages two main technological approaches: inductive power transfer and magnetic resonance charging. Inductive power transfer operates using electromagnetic fields to transfer energy between two coils: a transmitter coil embedded in the charging pad on the ground and a receiver coil installed on the aircraft. When an alternating current flows through the transmitter coil, it generates a fluctuating magnetic field that induces an electrical current in the receiver coil, which is then converted and used to charge the aircraft's batteries. The efficiency of this power transfer is highly dependent on the precise alignment and distance between the two coils, with modern systems achieving impressive efficiency rates through sophisticated engineering and control systems. Magnetic resonance technology represents a more advanced approach that uses resonant inductive coupling, where both the transmitter and receiver coils are tuned to the same electromagnetic frequency. This method enables efficient power transfer across larger air gaps and with greater tolerance for misalignment between the coils, providing significant advantages for aviation applications where precise positioning of an aircraft over a charging spot can be challenging. Ongoing research continues to optimize these technologies, with recent studies demonstrating groundbreaking designs for wireless magnetic coupling systems that achieve remarkable efficiency levels by addressing critical issues of leakage inductance and power loss.

Operational Benefits and Strategic Advantages

The implementation of wireless charging systems in electric aviation offers numerous strategic advantages that are crucial for making electric aircraft operations commercially viable and operationally efficient.

- **Enhanced Operational Efficiency:** Wireless charging enables automated charging processes that significantly reduce aircraft turnaround times compared to manual plug-in systems, which is essential for urban air mobility services requiring quick turnarounds to meet passenger demand.
- **Improved Safety Profile:** The elimination of physical connectors and cables reduces risks associated with wear and tear, electrical arcing, and short circuits, creating a safer working environment for ground crews and increasing overall system reliability in various weather conditions.
- **Automation Compatibility:** As a foundational technology for autonomous operations, wireless charging allows seamless integration with automated ground vehicles and self-parking aircraft, minimizing human intervention and streamlining operational processes.
- **Vehicle Design Optimization:** The potential for dynamic wireless charging during ground operations may allow for reduced battery sizes, decreasing overall aircraft weight and leading to improved payload capacity and flight efficiency.
- **Infrastructure Scalability:** Resonant charging systems can be designed to power multiple vehicle types from a shared charging network embedded in the apron, maximizing infrastructure utilization and return on investment.
- **Environmental Sustainability:** When integrated with smart grid technology and renewable energy sources, wireless charging facilities can optimize energy use, provide grid stabilization services, and further reduce the carbon footprint of aviation operations.
- **Reduced Maintenance Requirements:** Without physical connectors prone to mechanical damage from repeated use, wireless charging systems typically offer lower long-term maintenance costs and higher durability.
- **Future-Proofing Capability:** Embedded wireless charging infrastructure can often be upgraded as technology improves without requiring complete replacement, protecting long-term infrastructure investments.
- **Passenger Experience Enhancement:** For travelers, wireless charging translates to reduced waiting times and more frequent flight departures, making electric air travel more convenient and attractive.
- **High-Tempo Operation Support:** The high-power capabilities of modern wireless chargers are critical for supporting the intense operational

tempo of commercial eVTOL fleets, enabling practical charge cycles compatible with commercial service requirements.

Current Implementation and Industry Progress

Wireless charging technology is rapidly advancing from laboratory research to real-world aviation applications, with several pioneering projects demonstrating its practical feasibility and operational benefits. On conventional airport aprons, wireless charging systems are being implemented for ground support equipment, including baggage tugs, passenger buses, and catering trucks. These applications serve as valuable testbeds for the technology while providing immediate operational benefits through reduced emissions and improved efficiency.

The design of dynamic inductive charging networks for airport aprons represents a particularly promising development, with systems that embed inductive transmitter units in the road surface to charge electric vehicles while they perform their regular duties, effectively eliminating charging downtime entirely. In the emerging eVTOL sector, stationary wireless charging facilities are becoming the dominant solution for vertiport operations, with high-power units capable of delivering the rapid charging cycles required for commercial urban air mobility services. The market for these charging facilities is experiencing exponential growth, reflecting the aviation industry's confidence in wireless charging as a key enabling technology. Demonstration projects worldwide are validating the technology under real-world conditions, with several electric aircraft manufacturers partnering with infrastructure providers to develop integrated aircraft and charging solutions. These initiatives highlight the critical role of public-private partnerships in building the infrastructure ecosystem that will support both current and future wireless charging networks for electric aircraft.

Technical Challenges and Development Hurdles

Despite its significant potential, the widespread implementation of wireless charging in aviation faces several substantial challenges that require continued research and development effort. The high initial investment required for installing wireless charging infrastructure represents a major barrier to adoption, as the systems involve embedding transmitter coils in concrete or asphalt and installing high-frequency power supply units, all of which contribute to significant capital expenditure.

This financial challenge must be addressed through technological advancements, manufacturing scale, and innovative business models that distribute costs across multiple stakeholders. Technical challenges also remain, particularly regarding efficiency optimization across varying operating conditions. While wireless power transfer systems can achieve high efficiency rates under ideal conditions, performance can degrade with distance increases or coil misalignment, presenting engineering challenges for aviation applications where precise positioning may be difficult to maintain consistently. Managing electromagnetic interference in the dense electronic environment of airports and aviation facilities is another critical concern, as wireless charging systems must operate without disrupting sensitive navigation, communication, and safety equipment. Thermal management presents additional engineering challenges, as high-power wireless charging generates significant heat in both transmitter and receiver components, requiring sophisticated cooling systems to maintain efficiency, ensure safety, and prolong system lifespan. These technical hurdles, while substantial, are the focus of intensive research and development efforts across industry and academia.

Regulatory Framework and Standardization Needs

The integration of wireless charging technology into aviation operations requires the development of comprehensive regulatory frameworks and industry standards to ensure safety, reliability, and interoperability. Aviation authorities worldwide are beginning to address the certification requirements for wireless charging systems, though the regulatory landscape remains in early stages of development. Safety certification represents a particularly complex challenge, as wireless charging systems introduce new failure modes and potential hazards that differ from conventional conductive charging systems. Regulatory agencies must establish rigorous testing protocols and performance standards to ensure that wireless charging systems can withstand the demanding environmental conditions of aviation operations, including temperature extremes, vibration, moisture, and electromagnetic compatibility requirements.

The development of international standards is equally important for ensuring interoperability between different aircraft types and charging systems from various manufacturers. Standardization efforts must address key parameters including power levels, communication protocols, frequency bands, and physical interface characteristics to create a cohesive ecosystem that supports widespread adoption. Industry consortia and standards organizations are actively working to develop these specifications, though achieving global consensus will require extensive collaboration among aircraft manufacturers, infrastructure providers, regulatory agencies, and other stakeholders. The establishment of clear regulations and standards is essential for building confidence in wireless charging technology and facilitating the investment needed for large-scale deployment.

CHAPTER 3

Aircraft Design for Electrification

3.1 ARCHITECTURE OF ELECTRIC AIRCRAFT

3.1.1 Fully Electric vs. Hybrid-Electric Designs

Neither architecture is universally "better"; each dominates different parts of the mission map. Over the next decade, expect FE to expand in urban air mobility, training, and short regional cargo/passenger routes where high utilization and quiet operations yield strong economics. Hybrids will anchor longer regional connectivity, offering immediate fuel and noise reductions without waiting for grid upgrades or step-change battery density. As pack-level energy inches upward, thermal systems improve, and energy-aware flight operations mature, FE range will grow and HE batteries will shrink, converging toward optimized, mission-specific solutions. In the end, the winners will be operators who align architecture with mission physics and infrastructure reality, using data-rich energy management to turn electrons and fuel into reliable, affordable seat-miles.

Definitions and Core Architectures

Fully electric (FE) aircraft use batteries as the sole onboard energy store, driving electric motors through inverters. They can be fixed-wing or vertical-lift, with distributed propulsion common for efficiency and control. Their architecture centers on high-voltage DC buses, battery management, and thermal systems. Because batteries must cover both peak power and total energy, FE airframes are designed to minimize power-hungry segments—especially hover—and to achieve high lift-to-drag ratios in cruise. The pack is a structural and financial centerpiece, influencing aerodynamics, center of gravity, and business model.

Hybrid-electric (HE) comes in flavors. Series hybrids use an engine-driven generator to supply a high-voltage bus, with batteries supplementing for takeoff and contingencies; propulsors are all electric. Parallel hybrids mechanically couple the engine to the propulsor and layer an electric machine in the drive, enabling engine-off segments or electric assist at high demand. Turbo-electric variants are series hybrids using a turbogenerator sized for cruise, relying on batteries for climb or vertical segments. Series architectures simplify propulsor distribution and enable advanced control, while parallel systems ease retrofits of existing airframes. Each choice trades efficiency, redundancy, mass, and certification complexity differently.

Energy and Performance Envelope

Energy density sets the rules. Pack-level gravimetric energy in certifiable aviation batteries today typically ranges around 180–260 Wh/kg, whereas jet fuel contains orders of magnitude more chemical energy but converts to shaft power at lower efficiency. FE aircraft thrive on missions where total energy demand is modest—short stage lengths, early transition to wing-borne flight, and predictable weather. Their performance is often power-limited during takeoff, hover, or go-around, so designers favor large rotors (low disk loading), efficient wings, and tight thermal control to keep C-rates in bounds. HE shifts the constraint. With a generator providing cruise power, batteries can be sized primarily for peaks and reserves, not for the entire mission energy. That improves effective payload-range and reduces sensitivity to headwinds or diversions. The generator's efficiency map matters: running a turbogenerator near its "best efficiency island" can yield significant fuel savings versus throttle-chasing a conventional engine across varying loads. However, added mass from the engine, generator, and fuel system offsets some gains, and thermal management becomes a two-front war: keeping both batteries and hot-section components within limits.

Mass, Complexity, and Reliability

FE designs can be mechanically simpler: no gearboxes, fuel systems, or hot rotating machinery. Simpler hardware often translates to lower maintenance hours and fewer wear-out parts. Reliability, however, still depends on battery health, inverter robustness, cooling, and software. Redundancy in distributed propulsion can offer graceful degradation if a motor or inverter fails, assuming electrical and thermal margins are designed accordingly.

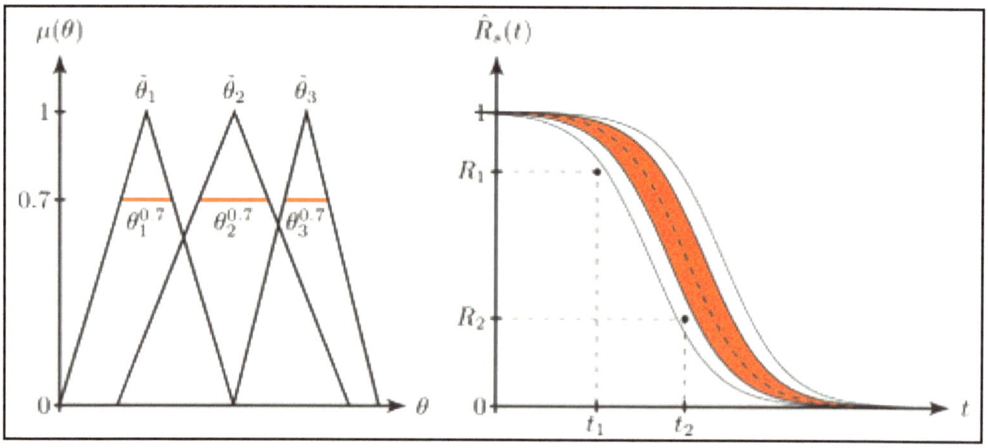

Strengths of Hybrid Architectural Design

HE introduces complexity and redundancy through diversity. An engine and generator add components, plumbing, and controls, but they also supply alternative energy paths. If the battery module faults, the generator can sustain essential loads; if the generator fails, the battery can cover a safe landing in many mission profiles. The trade is weight and maintenance: hot sections require inspections, fuel handling procedures return, and the energy management system must coordinate multiple sources under failure modes. Certification authorities scrutinize that coordination, demanding fail-operational behavior for critical functions.

Infrastructure and Turnaround

On the ground, FE operations depend on high-power charging or battery swapping. Fast charging reduces turn time but accelerates aging and raises thermal demands; slower charging preserves cycle life but constrains utilization. Airports and vertiports need megawatt-class capacity for fleets, plus buffers to avoid grid peaks. Battery swapping decouples utilization from charge rate but adds mass, connectors, and logistics complexity.

HE eases infrastructure pressure. A smaller battery can be charged at lower power or even off-peak, while conventional refueling covers most energy needs. That smooths adoption at regional airports with limited grid capacity and can raise dispatch reliability in extreme weather. However, dual-energy operations mean handling both fuel and electricity, training ground crews in additional procedures, and maintaining safe interfaces for charging and refueling on tight ramps.

Economics and Total Cost of Ownership

Operating cost in FE aircraft is driven by electricity price, battery depreciation, and minimal scheduled maintenance. Electric motors and inverters boast high efficiency and few moving parts; fewer fluids and overhauls reduce direct maintenance cost per flight hour. The big economic variable is battery life: depth of discharge, fast-charge frequency, temperature exposure, and C-rate peaks determine replacement cadence and residual value. High utilization on short routes can amortize packs efficiently if charging is well managed.

HE economics are a balancing act. Fuel burn drops versus conventional aircraft—often materially on short sectors—while the battery can be smaller and live longer than in FE, reducing depreciation risk. But engines, generators, and dual systems increase capital and maintenance costs. The best cases keep the generator near optimal efficiency in cruise and use the battery strategically for climb and contingencies, avoiding perpetual charge–discharge cycling that erodes life. For thin regional routes underserved today, HE can unlock viable service with modest infrastructure investment, narrowing the cost gap to FE while retaining longer range and flexibility.

Environmental and Community Impacts

FE aircraft deliver zero tailpipe emissions and very low noise, especially during takeoff in urban corridors. Lifecycle emissions depend on grid carbon intensity and battery materials, but as grids decarbonize, FE's climate advantage grows. Quiet operations expand operating windows and improve community acceptance—critical for urban air mobility and airports with strict noise abatement. HE reduces emissions and noise relative to conventional aircraft by offloading peaks to the battery and allowing smaller, more efficient engines. It cannot match FE's local air quality and noise benefits, but the gap shrinks on longer routes where FE would require large, heavy packs. Sustainable aviation fuel (SAF) can further lower HE's lifecycle carbon intensity, offering a pragmatic path while battery technology matures.

Certification and Safety Considerations

Certification landscapes differ. FE programs focus on battery safety (containment, non-propagation, venting), high-voltage protection, thermal management, and energy-reserve compliance. Demonstrating accurate state-of-energy estimation and robust emergency margins is central. HE adds classical engine/turbine certification elements and power split logic that must remain safe under faults. Software and hardware assurance follow standards like DO-178C and DO-254 at appropriate Design Assurance Levels, while battery systems meet DO-311A. For eVTOLs, special conditions (e.g., EASA SC-VTOL) shape reserve rules—often expressed as hover minutes or alternate profiles—affecting both FE and HE mission planning.

Each Shines

FE is compelling for short, high-frequency missions: urban air mobility corridors, training flights, island hops, short regional sectors with reliable charging, and operations where quiet, zero-emission flight is valued. These missions exploit FE's strengths—low energy per mission, quick cycles, and minimal maintenance—while minimizing exposure to long reserves or extreme weather variability. Fixed-wing FE trainers are already in service, and several cargo and passenger demonstrators show promise on 100–300 km routes where high lift-to-drag and efficient operations keep energy budgets in check.

HE excels where range, reserves, and infrastructure constraints dominate. Regional routes spanning 200–800 km with variable winds and alternates, operations from hot-and-high airports, and markets with limited grid access favor hybrids. Turbogenerator-based series hybrids support distributed propulsion and short-field performance while preserving payload and schedule reliability. Parallel hybrids offer pragmatic retrofits for legacy fleets, using electric assist to reduce fuel in climb and enabling partial-electric taxi or approach to cut noise near communities.

Decision Criteria for Choosing Fully Electric vs. Hybrid-Electric

- **Mission Length and Variability:** Short, predictable sectors favor FE; longer routes with headwind uncertainty or demanding alternates push toward HE.
- **Power Profile:** High peak-to-average power (hover, steep climb) strains FE packs; HE can cover peaks with batteries and sustain cruise on a generator.
- **Infrastructure Readiness:** Megawatt-class charging availability enables FE fleet ops; limited grid capacity or remote bases tip the scale to HE.
- **Dispatch Reliability Needs:** HE's onboard generation cushions weather and ATC delays; FE is more sensitive to reserve margins and temperature derates.
- **Noise and Emissions Priorities:** FE minimizes both at the point of use; HE reduces but cannot eliminate them, though SAF can mitigate lifecycle CO_2.
- **Total Cost of Ownership:** FE cuts maintenance and energy costs but budgets for battery depreciation; HE balances fuel, maintenance, and smaller-battery life.
- **Certification Path and Timeline:** FE avoids engine certification but faces stringent battery rules; HE must certify both engine and electric subsystems.
- **Airframe and Payload Constraints:** FE packs consume weight/volume; HE adds engine mass but often preserves payload-range on regional missions.
- **Fleet Strategy and Scalability:** FE shines in dense corridors with standardized vertiports; HE scales faster across diverse airports with existing fuel logistics.
- **Technology Risk Appetite:** FE bets on rapid battery improvements and grid build-out; HE hedges with mature engines while adopting electric benefits incrementally.

3.2 WEIGHT AND THERMAL MANAGEMENT CONSIDERATIONS

Weight and thermal management are not competing priorities—they are interconnected elements of a single design challenge. For engineers, success lies in moving beyond siloed optimization to integrated planning: evaluating every weight reduction for its thermal impact, and every thermal solution for its mass penalty. The ten critical considerations outlined here provide a practical framework for this work, while emerging technologies are breaking long-standing tradeoffs. In an era where efficiency and sustainability are paramount, mastering this balance is not just a design choice—it is a prerequisite for competitive, reliable, and safe engineering solutions.

3.2.1 Interdependent Challenge: Weight Reduction vs. Thermal Efficiency

At the heart of weight and thermal management is a fundamental tension: the materials and designs that minimize mass often compromise thermal performance, and vice versa. Consider the example of EV battery packs, which represent 25-30% of an EV's total weight. To extend range, manufacturers increasingly use lightweight aluminum enclosures instead of steel—reducing mass by 30-40%—but aluminum has lower thermal mass than steel, meaning it cannot absorb as much heat during rapid charging or high-load driving.

This requires additional thermal solutions, such as phase change materials (PCMs) or thin cooling channels, to prevent battery overheating. In aerospace, the tradeoff is even starker: satellite payloads must be as light as possible to minimize launch costs, but sensitive electronics like imaging sensors generate heat that can distort data or damage components. A single kilogram added to a satellite can cost $10,000 or more to launch, so engineers cannot afford to oversize thermal systems—yet undersizing them risks mission failure. This tension is not limited to high-tech industries: in data centers, dense server racks reduce floor space but trap more heat, requiring heavier cooling systems to compensate. The key insight here is that weight and thermal management are not independent variables; every decision to reduce mass must be evaluated for its impact on heat transfer, and every thermal solution must be optimized for minimal weight.

3.2.2 Industry-Specific Imperatives for Weight and Thermal Management

The priorities for weight and thermal management vary significantly by industry, driven by unique operational constraints and performance goals. Three key sectors where this balance is particularly critical include:

Automotive (Electric Vehicles)

For EVs, range anxiety and charging speed are primary consumer concerns—both directly tied to weight and thermal management. Battery packs are the heaviest component, so lightweighting is a top priority, but poor thermal control accelerates battery degradation: a 10°C increase in operating temperature can cut battery lifespan by 50%. To resolve this, manufacturers are moving beyond standalone thermal systems to integrated designs: Tesla's 4680 battery cells use a structural pack that doubles as a heat sink, eliminating redundant weight while improving heat distribution. This approach reduces total pack weight by 10% while cutting thermal management costs by 20%.

Aerospace

In aerospace, every gram matters. For launch vehicles, the cost of lifting mass to low Earth orbit is approximately 2,700perkilogram—rising to 2,700 perkilogram—rising to 50,000 per kilogram for deep space missions. Yet thermal challenges are extreme: re-entry vehicles like SpaceX's Dragon capsule face temperatures of up to 1,600°C, requiring thermal protection systems (TPS) that add significant mass. NASA's Space Shuttle TPS weighed over 3,000 kg; engineers had to balance tile thickness (for thermal resistance) with mass (to keep launch costs manageable). Modern TPS materials, such as silica aerogels and carbon-carbon composites, offer better thermal resistance per unit mass, reducing TPS weight by 40% compared to Shuttle-era designs.

Aviation and Space Technology

Data Centers

Data centers consume 2% of global electricity—much of it for cooling—and are under pressure to reduce both energy use and physical footprint. Dense server racks pack more computing power into smaller spaces but generate more heat, requiring heavier cooling systems. To balance this, operators are adopting liquid cooling systems that are 30% more energy-efficient than air cooling and add less weight per server than traditional fans. Lightweight server chassis made from magnesium alloys also reduce structural floor load, allowing more racks per room without costly upgrades.

3.2.3 Core Design Considerations to Balance Weight and Thermal Performance

To navigate the tradeoff between weight reduction and thermal control, engineers must prioritize a set of interconnected design principles grounded in material science, computational modeling, and real-world operational data. These critical considerations provide a framework for making informed tradeoffs:

- **Material Thermal Conductivity-to-Weight Ratio:** Prioritizing materials (e.g., aluminum-silicon alloys) that offer high thermal transfer per unit mass, rather than just high conductivity or low weight in isolation. For example, pure copper has excellent conductivity but is dense; aluminum-silicon alloys have 60% of copper's conductivity but only 30% of its density, making them a more weight-efficient choice for EV battery heat sinks.
- **Topology Optimization for Thermal Components:** Using computational design to remove excess material from heat sinks or cooling channels while maintaining heat dissipation capacity. Advanced finite element analysis (FEA) tools can identify non-critical sections of a heat sink, allowing engineers to slim them down—reducing weight by up to 20% without sacrificing performance.
- **Passive vs. Active Thermal Systems Tradeoffs:** Passive systems (heat pipes, PCMs) add less weight than active systems (pumps, fans) but may be less effective in extreme conditions. In satellite design, heat pipes are preferred over active cooling loops because they have no moving parts and add minimal mass, even though they rely on temperature gradients to function.
- **Thermal Mass Optimization:** Balancing the need for a thermal buffer (to absorb temperature spikes) with the weight penalty of high-mass materials. In EV rapid charging, a thin layer of PCM can absorb excess heat without adding significant weight—unlike a heavy metal heat sink that would provide similar buffering but reduce range.
- **Integrated Component Design:** Combining structural and thermal functions (e.g., EV battery enclosures that double as heat sinks) to avoid redundant weight. Traditional enclosures are purely structural, but modern cast aluminum designs with integrated cooling channels eliminate separate heat sinks, reducing total system weight by 15-20%.

- **Operational Load Profiling:** Tailoring thermal solutions to real-world usage (e.g., EVs for short commutes vs. long hauls) to avoid over-engineering weight. For urban commuter EVs, a simple passive cooling system may suffice, whereas long-haul trucks require robust active cooling—but engineers can size systems to match actual loads, not theoretical maxima.
- **Coating and Surface Treatment Efficiency:** Using thin, lightweight thermal coatings (e.g., ceramic thermal barriers) instead of thick material layers to reduce heat transfer without added mass. In aerospace gas turbines, 0.1mm-thick ceramic coatings can lower metal component temperatures by 200°C, eliminating the need for thicker, heavier superalloy sections.
- **Weight Distribution Impacts on Thermal Performance:** Ensuring that lightweighting efforts do not disrupt airflow or heat dissipation pathways. For example, removing material from an aircraft engine nacelle to reduce weight may alter airflow around the turbine, increasing heat buildup—so engineers must model airflow changes alongside weight reductions.
- **Lifecycle Weight-Thermal** tradeoffs: Considering long-term costs (e.g., battery replacement due to poor thermal management) against short-term weight savings. A 5% weight reduction in an EV battery pack that leads to 20% faster degradation is not cost-effective over the vehicle's lifespan, as replacement costs far outweigh initial range gains.
- **Regulatory and Safety Constraints:** Aligning weight reductions with thermal safety standards (e.g., EV battery thermal runaway prevention) to avoid compromising compliance. Some lightweight battery enclosures may not meet fire resistance standards, so engineers must balance material weight with required thermal insulation properties.

3.2.4 Emerging Technologies to Optimize the Weight-Thermal Balance

As industries push for greater efficiency, new technologies are expanding the boundaries of what is possible in weight and thermal management. Advanced nanocomposite PCMs, for example, have higher thermal storage capacity per unit mass than traditional PCMs—graphene-reinforced variants can store 20% more heat while being 15% lighter, making them ideal for EV battery packs.

Additive manufacturing (3D printing) is another game-changer. By printing thermal components like heat sinks with complex, optimized geometries, engineers can remove up to 30% of excess material while improving heat dissipation. 3D-printed aluminum heat sinks for satellite electronics have reduced weight by 25% while increasing thermal efficiency by 18% compared to machined designs. Finally, graphene-based thermal materials offer exceptional conductivity at minimal weight. Graphene has 100 times the thermal conductivity of copper but is 1,000 times lighter, making it ideal for thin thermal interfaces in electronics. While mass production remains costly, early applications in high-end smartphones and satellite sensors have demonstrated its potential to cut thermal system weight by up to 40%.

3.3 AERODYNAMIC OPTIMIZATIONS FOR ELECTRIC PROPULSION

Aerodynamic optimization is a cornerstone of electric aviation, enabling aircraft to achieve maximum efficiency, safety, and environmental performance. Distributed electric propulsion, high-aspect-ratio wings, adaptive airfoils, streamlined fuselages, and noise reduction strategies collectively contribute to improved lift-to-drag ratios, extended range, and lower energy consumption. Electric aircraft present unique opportunities for innovative aerodynamic solutions, but they also introduce challenges related to weight, thermal management, and structural integration.

By addressing these challenges through sophisticated modeling, testing, and design strategies, engineers can ensure that electric propulsion reaches its full potential. Electric aviation depends on holistic aerodynamic optimization, integrating propulsion, structure, and thermal systems. With continued research and technological advancements, electric aircraft will achieve safer, quieter, and more efficient flight, transforming regional, urban, and short-haul air transportation. Aerodynamic innovations, combined with energy-efficient propulsion, will define the next generation of sustainable aviation.

3.3.1 Aerodynamic Considerations in Electric Aircraft

Aerodynamics plays a central role in all aircraft, affecting lift, drag, stability, and control. Electric propulsion adds new dimensions to aerodynamic design. Unlike conventional engines, electric motors are smaller, lighter, and capable of being distributed across the airframe. This flexibility enables distributed electric propulsion (DEP), which can improve efficiency through optimized airflow management. However, designers must also address the weight and cooling requirements of batteries, structural adaptations for motor placement, and changes in the aircraft's center of gravity.

Optimizing aerodynamics for electric aircraft is not merely about reducing drag; it also involves enhancing lift-to-drag ratios, improving stability under variable loads, and ensuring efficient integration of electric propulsion systems. These considerations are critical because energy storage limitations make efficiency paramount for achieving practical flight ranges and payload capacities.

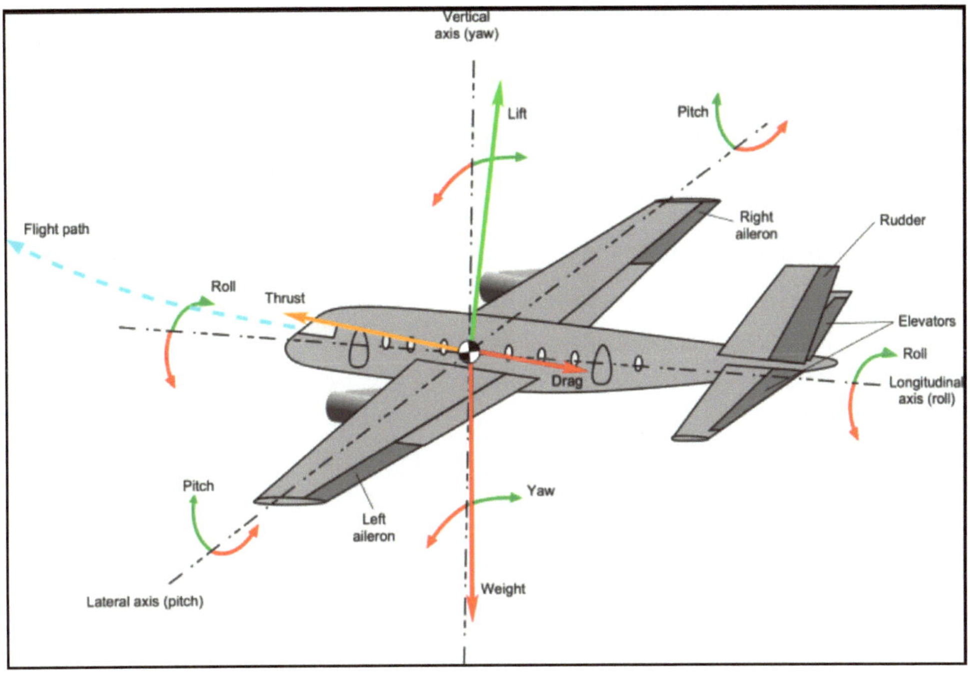

Equations of Motion for an Airplane, Aerospace Flight Vehicles

3.3.2 Distributed Electric Propulsion and Aerodynamic Benefits

One of the most promising advancements in electric aviation is distributed electric propulsion. DEP involves placing multiple small electric motors along the wings or fuselage, rather than relying on a few large engines. This configuration offers several aerodynamic benefits.

- First, DEP can enhance lift, particularly at low speeds. Motors with integrated propellers or fans can accelerate airflow over the wings, delaying stall and improving takeoff and landing performance. This is especially beneficial for regional aircraft, urban air mobility vehicles, and short-haul electric planes.
- Second, DEP reduces induced drag. By distributing thrust across multiple points, designers can manipulate airflow more effectively, creating smoother lift distribution along the wingspan. This not only improves energy efficiency but also enhances aircraft stability and reduces structural stress.
- Finally, DEP allows for thinner wings and more streamlined fuselages, as high-lift devices can be partially replaced by motor-assisted airflow control. Optimizing the placement of each motor requires sophisticated computational fluid dynamics (CFD) modeling to balance lift, drag, and thrust throughout the flight envelope.

3.3.3 Wing Design Optimizations

Wing design is crucial for aerodynamic efficiency in electric aircraft. Electric propulsion enables new approaches to aspect ratio, wing sweep, and airfoil selection. Higher aspect ratio wings, which are long and slender, reduce induced drag and improve lift-to-drag ratios, benefiting range and efficiency. However, long wings introduce structural weight challenges, which must be mitigated using lightweight composite materials. Airfoil selection is another key factor.

For electric aircraft, designers often prioritize airfoils that perform efficiently across a wide range of speeds, as electric motors can provide continuous thrust at low speeds. Adaptive or morphing wings are also being explored, allowing the airfoil shape to change dynamically in response to flight conditions. Such innovations can reduce drag during cruise while enhancing lift during takeoff and landing. Additionally, wingtip devices such as winglets or active wingtip fans can be integrated with electric propulsion systems. These devices reduce vortex-induced drag and can be combined with DEP to further optimize efficiency. CFD and wind tunnel testing are essential to identify the optimal wing configurations for specific electric aircraft designs.

3.3.4 Fuselage and Body Aerodynamics

The fuselage of electric aircraft also requires aerodynamic optimization. Electric propulsion often allows for more compact nacelles or engine housings, which reduces frontal area and form drag. Designers can adopt sleeker fuselage shapes with smoother contours, minimizing turbulence and improving laminar flow over the aircraft body. Battery placement within the fuselage introduces weight distribution challenges. Optimal aerodynamic performance requires careful attention to the center of gravity and mass distribution to prevent stability issues. Internal airflow management for cooling batteries and electronics also impacts fuselage design, as vents, ducts, and radiators must be integrated without increasing drag excessively.

3.3.5 Cooling and Thermal Management

Electric propulsion systems generate heat from both motors and batteries. Aerodynamic optimization must therefore account for thermal management. Airflow over cooling surfaces, ducts, and radiators must be efficiently designed to maintain system performance while minimizing drag penalties. Innovative approaches include integrating boundary layer cooling, using airflow along the fuselage and wings to dissipate heat, and employing passive cooling surfaces to reduce reliance on active fans or pumps. The integration of thermal management with aerodynamic design is critical for high-power operations such as takeoff and climb, where heat generation is maximal. Poorly managed cooling can increase drag, reduce efficiency, and compromise motor or battery longevity.

3.3.6 Noise Reduction through Aerodynamic Design

Electric propulsion is inherently quieter than conventional engines, but aerodynamic noise remains a concern, especially in urban air mobility applications. Propeller design, blade shape, and spacing interact with airflow over the wings and fuselage to influence noise levels. By optimizing propeller aerodynamics, blade tip clearance, and motor placement, designers can reduce both tonal and broadband noise. Additionally, streamlined fuselage shapes and smooth wing surfaces help minimize turbulent flow that contributes to noise. DEP systems can also be configured to operate at lower rotational speeds while maintaining thrust, further reducing noise during critical phases of flight such as takeoff and landing.

Table: Aerodynamic Optimization Features for Electric Aircraft

Feature	Optimization Purpose
Distributed Electric Propulsion	Enhances lift, reduces induced drag
High Aspect Ratio Wings	Improves lift-to-drag ratio and range
Adaptive Airfoils	Optimizes performance across different speeds
Wingtip Devices	Reduces vortex-induced drag
Sleek Fuselage	Minimizes form drag and turbulence
Compact Nacelles	Reduces frontal area and drag
Battery Placement	Maintains center of gravity and stability
Thermal Management Ducts	Manages heat without increasing drag
Low-Speed Propeller Optimization	Improves takeoff and landing efficiency
Noise-Reducing Blade Design	Minimizes acoustic impact in urban areas

3.3.7 Aerodynamic Optimization Strategies

- **Distributed Propulsion:** Placement of multiple small motors to optimize lift and reduce drag.
- **High Aspect Ratio Wings:** Long, slender wings to minimize induced drag.
- **Adaptive or Morphing Wings:** Dynamic airfoil shapes for varying flight conditions.
- **Optimized Winglets:** Reduce tip vortices and improve efficiency.
- **Streamlined Fuselage:** Minimizes form drag and promotes laminar flow.
- **Compact Motor Nacelles:** Reduces frontal area and drag contribution.
- **Weight Distribution Management:** Maintains center of gravity with battery placement.

- **Efficient Thermal Management:** Integrates cooling systems with minimal drag impact.
- **Noise-Optimized Propellers:** Blade design and placement reduce acoustic emissions.
- **Integrated CFD Modeling:** Uses simulation to predict and optimize aerodynamic performance.

These strategies collectively enhance the efficiency, performance, and environmental impact of electric aircraft while addressing the unique challenges of energy-limited propulsion systems.

3.3.8 Challenges and Trade-Offs

Aerodynamic optimization in electric aircraft involves complex trade-offs. High-aspect-ratio wings reduce drag but increase structural weight, necessitating lightweight materials. Distributed propulsion improves lift and reduces induced drag, but adds complexity to motor controls, wiring, and maintenance. Cooling systems are essential but may increase drag if poorly integrated. Electric aircraft also face range limitations due to battery energy density, which makes aerodynamic efficiency even more critical. Designers must balance low-speed performance for takeoff and landing with high-speed cruise efficiency. Multi-disciplinary optimization, combining aerodynamics, structures, propulsion, and thermal systems, is essential to achieving the best overall performance.

3.4 INTEGRATION OF BATTERIES AND PROPULSION SYSTEMS

The integration of batteries and propulsion systems represents a fundamental transformation in aircraft design that promises to redefine the future of aviation. While significant challenges remain, particularly regarding energy storage density and thermal management, the rapid pace of technological advancement suggests these barriers will be progressively addressed in the coming years. The benefits of electric propulsion—including reduced emissions, lower noise, and operational flexibility—provide strong motivation for continued investment and development. The evolution of these systems will likely follow a gradual path, beginning with smaller aircraft on shorter routes and expanding to larger aircraft and longer ranges as technology improves. Success will require continued collaboration across multiple disciplines, including materials science, electrical engineering, aerodynamics, and thermal management. The growing emphasis on sustainable aviation and the increasing maturity of electric propulsion technologies suggest that integrated battery and propulsion systems will play an increasingly important role in the aviation ecosystem. While electric propulsion may not completely replace conventional systems in the near term, it is poised to become a significant and growing segment of the aviation market, contributing to a more diverse, efficient, and environmentally responsible aviation industry. The continued development of these integrated systems represents not merely an incremental improvement but a fundamental step forward in aircraft technology with the potential to transform how we think about air transportation.

3.4.1 Core Propulsion System Architectures

Several technological pathways exist for integrating batteries with aircraft propulsion, each with distinct operational principles and applications. The choice of architecture involves critical trade-offs between range, power, complexity, and technological maturity. Fully electric systems represent the most direct approach, where multiple electric motors powered exclusively by on-board batteries drive all propulsors. This configuration offers zero operational emissions and simpler mechanical systems but faces limitations due to current battery energy density constraints. Hybrid turboelectric systems provide an alternative pathway, combining a gas turbine engine that drives a generator with battery systems that can provide peak power. This architecture maintains the ability to use sustainable aviation fuels while offering a practical balance between operational range and emissions reduction. Distributed electric propulsion represents perhaps the most radical departure from conventional aircraft design, employing numerous small electric motors and propulsors distributed across the airframe instead of relying on a few large engines. This approach enables novel aerodynamic efficiencies and provides built-in redundancy that enhances safety while enabling new aircraft configurations previously considered impractical.

3.4.2 Benefits of Integrated Systems

The tight coupling of batteries and propulsion unlocks several transformative advantages over traditional aircraft systems, leading to superior performance and environmental outcomes.

- **Significant Emissions Reduction**: Fully electric systems produce zero in-flight emissions, while hybrid systems can dramatically reduce fuel burn and carbon dioxide output compared to conventional aircraft propulsion systems.
- **Enhanced Operational Efficiency**: Integrated electric propulsion systems enable substantially lower operating costs through improved energy efficiency and reduced maintenance requirements compared to traditional turbine engines.
- **Noise Abatement**: Electric propulsion systems operate much more quietly than combustion engines, reducing community noise impact and enabling operations in urban environments where noise restrictions are stringent.

- **Shortened Takeoff and Landing**: Distributed electric propulsion creates aerodynamic benefits that increase lift at low speeds, allowing aircraft to operate from shorter runways and expanding potential operational locations.
- **Improved Safety through Redundancy**: Multiple independent electric motors provide inherent redundancy, ensuring that the failure of individual components does not compromise overall aircraft safety or controllability.
- **Propulsion-Based Aircraft Control**: Precise independent control of multiple motors enables new forms of aircraft control, reducing reliance on traditional control surfaces and creating more responsive handling characteristics.
- **Aerodynamic Synergies**: Strategic placement of electric propulsors can improve overall aerodynamics by re-energizing boundary layer airflow and reducing drag, leading to improved efficiency.
- **Flexible Power Management**: Hybrid systems allow intelligent power allocation between different energy sources, optimizing each component's operation for maximum overall efficiency across different flight phases.
- **Increased Design Freedom**: Electric propulsion liberates aircraft designers from traditional constraints, enabling more innovative airframe configurations optimized for specific missions and operational requirements.
- **Potential for Lower Infrastructure Dependency**: Electric and hybrid systems can reduce reliance on centralized fossil fuel infrastructure, increasing operational flexibility and energy resilience.

3.4.3 Critical Challenges and Technical Hurdles

Despite the promising benefits, the path to widespread adoption of integrated battery and propulsion systems faces substantial technical and operational challenges that require continued research and innovation. The most significant barrier remains the fundamental disparity in energy density between batteries and conventional aviation fuels.

Even advanced aviation batteries store only a small fraction of the energy per unit weight of kerosene, creating inherent limitations for aircraft range and payload capacity. This energy density challenge currently restricts practical electric aircraft applications to shorter routes and smaller aircraft categories, with longer-range electric flight awaiting future battery technology breakthroughs. Thermal management represents another critical engineering challenge, as electric motors and batteries generate significant heat during operation that must be effectively dissipated. The cooling systems required add considerable weight and complexity, directly impacting overall aircraft performance and efficiency.

The aviation certification process presents additional hurdles, as new technologies must undergo exceptionally rigorous and lengthy validation processes to ensure safety and reliability. These certification requirements extend development timelines and create substantial financial barriers, particularly for new market entrants. Supporting infrastructure represents a further challenge, as most airports lack the high-power charging capabilities needed for practical electric aircraft operations, requiring significant investment in ground-based systems to enable widespread adoption.

3.4.4 Thermal Management and System Integration

The thermal characteristics of integrated battery and propulsion systems present one of the most complex engineering challenges in electric aircraft development. Effective thermal management is crucial not only for maintaining system efficiency but also for ensuring safety and longevity of components. Battery systems particularly require precise temperature control, as both excessive heat and cold can degrade performance and potentially create hazardous conditions. Electric motors, while generally more efficient than combustion engines, still generate significant waste heat that must be dissipated, especially during high-power operations like takeoff and climb.

The integration of thermal management systems must be considered from the earliest design stages, as these systems can account for a substantial portion of the total propulsion system weight. Advanced cooling techniques, including liquid cooling systems and phase-change materials, are being developed to address these challenges while minimizing weight penalties. The interplay between thermal management, power density, and overall system efficiency requires sophisticated modeling and optimization to achieve practical aircraft designs. Furthermore, the thermal characteristics of these systems vary significantly across different flight phases and environmental conditions, requiring adaptive control strategies that can maintain optimal temperatures throughout the flight envelope. The successful integration of thermal management represents a critical enabling technology for practical electric aviation.

3.4.5 Certification and Safety Considerations

The certification process for integrated battery and propulsion systems introduces unique challenges that differ significantly from conventional aircraft systems. Aviation authorities worldwide are developing new frameworks to address the distinctive characteristics of electric propulsion, particularly regarding battery safety, electromagnetic compatibility, and system redundancy. Battery systems must demonstrate exceptional reliability and contain failure modes within acceptable safety parameters, requiring extensive testing and validation under various operating conditions and failure scenarios.

The certification process examines not only individual components but also their interactions within the complete integrated system, assessing how failures might propagate and affect overall aircraft safety. Distributed electric propulsion systems introduce additional certification considerations, as the failure of individual motors must not compromise the aircraft's continued safe flight and landing capability.

The electrical systems themselves must be designed to prevent electromagnetic interference with aircraft avionics and control systems, requiring careful shielding and system architecture. Furthermore, the maintenance and operational procedures for these new systems must be developed and validated to ensure continued airworthiness throughout the aircraft's service life. The evolving regulatory landscape for electric aircraft propulsion requires close collaboration between manufacturers and certification authorities to establish appropriate standards that ensure safety while enabling technological innovation.

3.5 NOISE REDUCTION AND ENVIRONMENTAL BENEFITS

Noise and environmental performance are now core design and business criteria, not afterthoughts. Electric and hybrid-electric aircraft unlock fundamentally quieter operations by eliminating combustion and enabling precise aerodynamic control, while also shrinking climate and airquality footprints. Yet realizing these benefits at scale demands system thinking: aerodynamic design paired with operational discipline, infrastructure that supports clean charging, and transparent measurement that earns public trust. As technology matures—better batteries, smarter energy management, and refined acoustic tooling—expect steady decibelbydecibel progress and lifecycle emissions that trend downward with the grid. The prize is substantial: more places served, more hours flown, and more communities that welcome aviation as a quiet, clean neighbor.

3.5.1 Noise Matters

Noise is not just a nuisance; it is a public health issue. Chronic exposure elevates stress hormones, disrupts sleep, impacts learning outcomes for children, and increases cardiovascular risk. The human ear is more sensitive to mid and highfrequency sounds, which is why Aweighted decibels (dBA) are commonly used. For aircraft certification, metrics such as Effective Perceived Noise in decibels (EPNdB), Sound Exposure Level (SEL), and maximum Aweighted level (Lmax) quantify both loudness and duration. Communities experience noise as events, not averages; a single loud overflight can define perceived annoyance more than a day's quiet. Therefore, reducing peak levels, shifting tonal content to less intrusive frequencies, and limiting the number of audible events are all meaningful strategies.

3.5.2 Acoustic Sources: Conventional vs. Electric

Conventional jets and turboprops produce noise from multiple sources: fan and compressor tones, turbulent jet exhaust, combustion rumble, and airframe noise from flaps, slats, and landing gear. Helicopters add bladevortex interaction (BVI) and tail rotor noise—sharp, impulsive signatures that carry far. In these aircraft, engine and exhaust noise often dominate during takeoff and climb, while airframe noise becomes more prominent on approach.

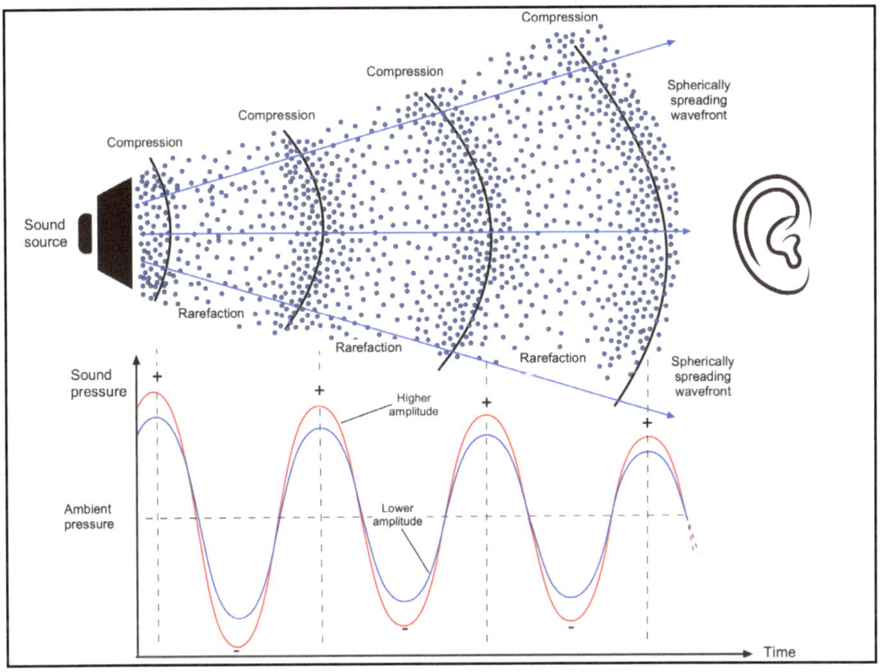

Sound Generation in Conventional and Electric Systems

Electric propulsion reshuffles this stack. With no combustion or exhaust plume, the signature shifts to aerodynamic and mechanical sources: rotor/propeller loading noise, tipvortex and broadband turbulence, motorgearbox tones (if a gearbox exists), and inverter switching harmonics. Because electric motors can operate at optimized speeds with precise control, designers have new freedom to shape the acoustic footprint—choosing rotor diameter, blade count, and RPM to trade total thrust for lower, less intrusive noise.

3.5.3 Electric Designs Change the Soundscape

Electric aircraft typically rely on larger-diameter rotors turning more slowly to produce lift efficiently. Low disk loading reduces the induced power required and the associated acoustic footprint. Distributed propulsion—multiple smaller propulsors—spreads thrust and allows finer control of local flow, reducing peak blade loading and harmonic content. Without combustor or jet noise, overall levels drop substantially at close range, and the spectrum shifts from harsh, highenergy tones to softer broadband noise.

This does not mean all electric aircraft are inherently quiet everywhere. Hover and lowspeed flight can still be loud, especially if rotors run at high tip speeds or interact with their own wakes or structures. Ducted fans can mask tip noise and permit tighter integration, but they risk tonal resonances unless treated with liners and carefully tuned spacing. Smart control helps: variable RPM and phase control can decorrelate tonal peaks, spreading energy and reducing perceived loudness without sacrificing performance.

3.5.4 Design Strategies for Quieter Aircraft

The most effective noise reductions come from aerodynamic and structural choices baked in early. Lower tip speeds and sufficient rotor diameter are foundational; tip Mach number is a reliable predictor of broadband and compressibility noise. Blade geometry—sweep, taper, and twist—reduces thickness and loading noise, while serrated or porous trailing edges damp turbulent eddies. Uneven blade spacing and multiharmonic phase control avoid tonal stacking that the ear finds objectionable. For tail-sitters and tiltrotors, transitions through edgewise flight are tuned to avoid BVI hot spots.

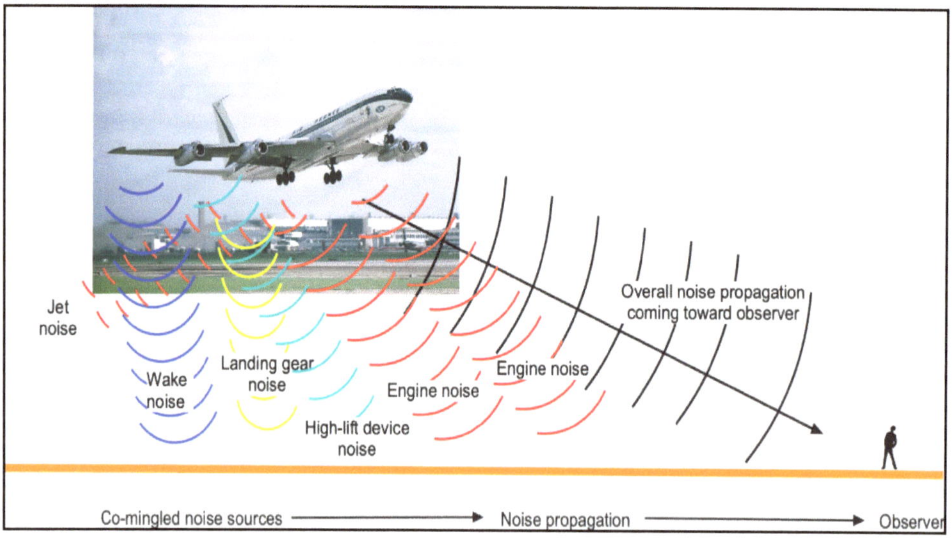

Noise-Minimizing Design Techniques in Aviation

Mechanically, directdrive motors eliminate gearbox whine and reduce maintenance. When gears are necessary, helical designs, precision grinding, and damping mounts suppress tones. Inverters with higher switching frequencies (enabled by

SiC or GaN devices) can push electrical noise above the most sensitive hearing bands, though care is needed to avoid increasing iron losses and EMI. Structural isolation—vibration mounts, composite layups with embedded damping—prevents transmission of residual noise into the fuselage and cabin.

3.5.5 Operations and Infrastructure that Cut Noise

Even a quiet aircraft can be loud if flown the wrong way. Continuous descent and energymanaged approaches reduce thrust changes and slat/gear deployment time. For eVTOLs, minimizing time in hover and transitioning to wingborne flight as early as safely possible reduces both noise and energy consumption. Route planning over corridors with higher ambient noise—highways, rivers, industrial zones—lowers audibility in residential areas. Vertiport design matters: siting pads away from reflecting surfaces, adding noise barriers or berms, and elevating pads above ground-level reflections all lower community exposure. Scheduling strategies—avoiding sensitive nighttime windows, staggering departures to prevent overlap—can keep event counts and SEL within acceptable thresholds. Pilot aids that visualize "noise rings" or realtime acoustic footprints help crews fly quieter profiles with confidence.

3.5.6 Environmental Benefits Beyond Noise

Electric and hybrid-electric aircraft bring significant environmental gains beyond acoustics. Eliminating tailpipe emissions removes nitrogen oxides (NOx), sulfur oxides (SOx), particulate matter, carbon monoxide, and unburned hydrocarbons at the point of use—critical for air quality in dense urban settings and downwind of airports. Electric drivetrains are inherently efficient; propulsion chain efficiencies of 85–95% in cruise compare favorably with 25–40% for small turbines and piston engines, reducing total energy required for a given mission when airframe and operations are optimized. Climate impacts improve as well. Fully electric operations have no inflight CO_2 emissions, and lifecycle carbon intensity depends on the grid mix used for charging. As renewable penetration rises, grams of CO_2 per passengerkilometer fall steadily. Hybrids cut fuel burn and allow smaller, more efficient engines to run at steady, optimal loads, with batteries covering peaks; the result is lower CO_2 and reduced nonCO_2 effects like contrail formation. Electric aircraft also eliminate fuel spills, reduce ramp heat and odors, and sharply cut brake dust and tire wear by enabling regenerative or propulsive deceleration strategies in some designs.

3.5.7 Lifecycle View and Grid Interactions

True environmental accounting considers materials extraction, manufacturing, operations, and endoflife. Batteries carry embodied impacts: mining and refining of nickel, cobalt, lithium, and copper; cell and pack production; and transportation.

Aviationgrade packs often have longer service lives when operated with tight thermal control and conservative charge windows, spreading embodied emissions over more flight hours. Secondlife use in stationary storage and highrecovery recycling further lowers net impact.

Chemistries like LFP reduce reliance on scarce materials and improve recyclability, albeit at lower energy density. On the operations side, smart charging aligned with lowcarbon hours (or colocated renewables) lowers lifecycle emissions, and onsite storage buffers grid peaks. Fleet scheduling can prioritize aircraft with healthier batteries on longer routes to avoid overreserving energy that would increase charging demand during carbonintensive periods. Transparency is key: publishing welltowing carbon intensity and noise maps builds trust and guides continuous improvement.

3.5.8 Community and Economic Upsides

Quieter, cleaner operations translate into durable community benefits. Expanded operating windows—early mornings and late evenings previously curtailed by curfews—boost aircraft utilization without increasing annoyance, provided noise thresholds are respected. Reduced emissions near schools and hospitals improve public health outcomes and align with city climate goals.

Property values near airports and vertiports may stabilize or even rise if noise footprints shrink and flight paths are predictable. For operators and airports, environmental performance is a competitive differentiator. Lower noise fees, access to noisesensitive locations, and eligibility for green financing improve project economics. Tourism and business travel can expand where local ordinances previously restricted aircraft operations. Most importantly, social acceptance—the ability to maintain routes without political pushback—becomes an asset rooted in measurable noise and emissions outcomes.

3.5.9 Measuring, Certifying, and Communicating Impacts

Standards guide both design and public reporting. Traditional fixedwing and rotorcraft follow FAA/EASA Part 36 and related ICAO Annex 16 noise certification procedures; emerging eVTOLs are aligning under special conditions (e.g., EASA SCVTOL) that emphasize urban context and eventbased metrics. For emissions, ICAO CAEP standards cover conventional engines; electric and hybrid systems increasingly report lifecycle metrics consistent with ISO 14040/44. On the ground, ISO 1996 methods for environmental noise mapping help planners visualize exposure in terms residents understand. Communication should be continuous and granular: noise contours updated with real operations, dashboards showing SEL distributions and complaint heat maps, and carbon intensity per route. Citizen engagement—listening sessions, transparent incident logs, and pilot briefings informed by community data—builds credibility. When residents can see that a new approach profile shaved 3 dB from Lmax at a school zone, or that charging shifted to solar hours cut route emissions by half, support becomes evidencebased, not speculative.

3.5.10 Practical Levers for Quieter, Cleaner Flight

- Use low disk loading with larger-diameter, slower rotors to reduce induced power and broadband noise, especially during hover and takeoff.
- Cap tip speed and optimize blade geometry (sweep, twist, serrations) to curb compressibility and trailingedge noise without sacrificing thrust.
- Favor directdrive motors or ultraquiet geartrains, and tune inverter switching with SiC/GaN devices to push tonal content outside sensitive bands.
- Employ distributed electric propulsion to lower perpropulsor loading, decorrelate tonal peaks, and enable quieter transition and approach control.
- Design continuous descent and early transition profiles that minimize time in highpower, highnoise flight regimes while protecting reserves.
- Site vertiports and approach/departure corridors over existing noise corridors and add physical barriers or elevation to limit reflections.
- Schedule operations to avoid sensitive hours and stagger departures to prevent overlapping events that inflate SEL in neighborhoods.
- Power charging with lowcarbon sources or offpeak tariffs, and precondition batteries thermally to reduce inflight derates and total energy draw.
- Choose battery chemistries and pack designs with high recyclability and plan for secondlife storage to reduce lifecycle environmental impact.
- Publish realtime noise and emissions data per route, and incorporate community feedback into flight procedures and infrastructure siting.

CHAPTER 4

Categories of Electric Aircraft

4.1 GENERAL AVIATION (LIGHT AIRCRAFT)

General aviation light aircraft are far more than a recreational pastime—they are a critical pillar of global aviation, supporting everything from pilot training to rural medical care and agricultural productivity. Their unique design and operational constraints reflect a need for resilience and versatility, traits that make them irreplaceable in environments where commercial aviation cannot reach.

While challenges like rising costs and a pilot shortage threaten their future, innovations in electric propulsion, composite materials, and training technology are opening new possibilities. For policymakers and industry leaders, supporting GA light aircraft is not just about preserving a hobby—it is about maintaining a resilient aviation ecosystem that serves every corner of the globe. As the next generation of electric light aircraft takes to the skies, the role of GA in connecting communities and powering economic activity will only grow more vital.

4.1.1 Defining Light Aircraft in General Aviation: Scope and Scale

The FAA's 12,500-pound MTOW threshold is a critical dividing line: above this, aircraft are classified as medium or heavy, subject to stricter operational and maintenance rules. Below it, light aircraft include everything from two-seat ultralights and light sport aircraft (LSA) to four-seat piston-powered trainers and small turboprops. The most iconic light aircraft—the Cessna 172—has been in production since 1956, with over 44,000 units built; it remains the global standard for flight training, with an estimated 70% of professional pilots logging their first solo hours in a 172.

Globally, there are over 350,000 registered light aircraft, with the U.S. accounting for nearly 60% of that total. What makes these aircraft unique is their versatility: unlike commercial jets, which are designed for high-capacity, high-altitude travel, light aircraft are optimized for slow speeds, short runways, and close proximity to the environments they serve. A light turboprop can land on a dirt strip in the Amazon rainforest; a glider can soar over mountain ranges for environmental surveying; and a two-seat LSA can provide affordable access to flight for hobbyists and new pilots alike.

4.1.2 Unsung Utility of Light Aircraft: Beyond Recreational Flight

For many, GA light aircraft evoke images of weekend hobbyists flying to small airshows—but this is a tiny fraction of their actual use.

- **The Largest Segment of GA Light Aircraft Operations is Flight Training:** every year, over 100,000 new pilots earn their private or commercial certificates in light aircraft, a pipeline that supplies 90% of the world's airline pilots. Beyond training, light aircraft are indispensable for utility tasks that commercial aviation cannot support. In agricultural aviation (or "ag aviation"), light turboprops like the Air Tractor 502 are used to apply pesticides, fertilizers, and seed to over 200 million acres of farmland in the U.S. alone—reducing labor costs by 70% and minimizing chemical waste compared to ground applications.
- **In Remote Regions, Light Aircraft are Lifelines:** in Alaska, where only 20% of communities have road access, GA light aircraft conduct 95% of medical evacuations ("medevacs"), transporting patients to urban hospitals in a fraction of the time it would take by boat or snowmobile. Even in densely populated areas, light aircraft support aerial surveying for pipeline inspections, wildfire detection, and search-and-rescue missions—their ability to loiter at low altitudes and land on unimproved runways makes them far more effective than helicopters for long-duration tasks.

4.1.3 Critical Operational and Design Considerations for Light Aircraft

Unlike commercial jets, which rely on teams of engineers, dispatchers, and maintenance crews to manage operations, light aircraft are often operated by single pilots with limited support. This means every design and operational choice is shaped by a need for simplicity, reliability, and margin for error. Below are ten critical considerations that define light aircraft operations:

- **Weight and Balance Precision:** Unlike commercial jets with dedicated weight specialists, GA pilots calculate load distribution themselves. Even a 50-pound shift in cargo or passengers can move the aircraft's center of gravity outside safe limits, increasing stall speed and reducing maneuverability—fatal in low-altitude flight.
- **Piston Engine Reliability Constraints:** Most light aircraft use four- or six-cylinder piston engines, which are simpler than turbofans but require far more frequent maintenance. The FAA mandates a "time between overhauls" (TBO) of 1,200–2,000 flight hours for most piston engines, compared to 10,000+ hours for commercial jet engines, to prevent in-flight failures.

- **Enhanced Weather Sensitivity:** Light aircraft have small wingspans and low cruise speeds (100–180 knots), making them far more vulnerable to turbulence, crosswinds, and thunderstorms than commercial jets. A 20-knot crosswind trivial for a Boeing 737 can be dangerous for a Cessna 172, so GA pilots must prioritize pre-flight weather briefings and contingency diversion plans.
- **Short-Field Performance Optimization:** Many light aircraft are designed to operate from runways as short as 1,500 feet—including unpaved grass or dirt strips. This requires specialized takeoff/landing techniques (e.g., full flaps for shorter rollouts) and reinforced landing gear to withstand rough surfaces.
- **Fuel Efficiency Tradeoffs:** Light aircraft are remarkably fuel-efficient per passenger—a Cessna 172 uses 8–10 gallons of fuel per hour, compared to 5 gallons per passenger per hour for a commercial jet. However, small fuel tanks limit range to 400–600 miles, so pilots must plan routes around small airports with fuel availability, especially in remote areas.
- **Regulatory Compliance for Private Operators:** GA pilots face stricter personal compliance rules than commercial crews. U.S. private pilots, for example, must complete a biennial flight review (1 hour of ground school + 1 hour of flight checks) every two years and maintain a current medical certificate with strict limits for conditions like high blood pressure.
- **Avionics Integration Balance:** Modern light aircraft often include glass cockpits (digital displays replacing analog gauges) for better navigation. But overloading with electronics adds weight and complexity; many pilots opt for hybrid systems (digital navigation + analog backups) to preserve payload capacity.
- **Emergency Procedure Preparedness:** Light aircraft lack commercial jets' redundant systems—most have one engine, one electrical system, and no pressurization. Pilots must master forced landing techniques (e.g., identifying flat fields) and carry emergency gear like locator transmitters and first-aid kits at all times.
- **Payload Flexibility vs. Safety:** Light aircraft typically carry 2–4 passengers or 500–1,000 pounds of cargo, but exceeding payload limits reduces climb rate and increases stall risk. A Cessna 172's maximum payload is around 800 pounds—including fuel, passengers, and luggage—so pilots must weigh every item to stay within limits.
- **Maintenance Accessibility:** In remote areas, light aircraft rely on local mechanics with limited tools. Manufacturers design engines and

airframes for easy field repairs—for example, quick-access cowlings that let pilots check oil levels or replace spark plugs without specialized equipment.

These considerations highlight a core truth about light aircraft: they are built for resilience, not speed or capacity. A commercial jet pilot can rely on automated systems and ground support; a GA pilot in the outback must troubleshoot engine issues or adjust load distribution without outside help.

4.1.4 Challenges Facing General Aviation Light Aircraft Today

For all their utility, GA light aircraft face significant headwinds that threaten their viability. The most pressing is rising operational costs: aviation gasoline (avgas)—the primary fuel for piston engines—has increased by 60% since 2018, and maintenance costs have risen by 30% due to supply chain shortages. Insurance premiums for GA pilots have also spiked: a private pilot flying a Cessna 172 now pays 1,500–1,500–3,000 annually, up from 800–800–1,200 a decade ago.

Equally concerning is the pilot shortage: the FAA projects the U.S. will need 18,000 new GA pilots annually over the next decade to replace retirees, but new private pilot certificates have dropped by 25% since 2010. This shortage stems from high training costs—10,000–10,000–15,000 for a private license—and a lack of young people entering the field. Regulatory burdens also slow innovation: overly complex certification processes for new light aircraft increase costs and delay the rollout of sustainable technologies. Finally, noise concerns in residential areas have led to restrictions on small airports, reducing GA access in densely populated regions.

4.2 REGIONAL AND SHORT-HAUL ELECTRIC PLANES

Light aircraft are a cornerstone of general aviation, providing essential services in training, recreation, emergency response, and aerial work. Their affordability, operational flexibility, and technological adaptability make them vital to the aviation ecosystem. Advances in electric propulsion, aerodynamics, composite materials, and avionics are transforming light aircraft, improving efficiency, safety, and environmental performance.

Despite challenges related to weather, fuel, infrastructure, and regulatory compliance, light aircraft continue to support economic activity and technological innovation. They serve as a bridge between recreational flying and commercial aviation, offering a platform for pilot development, experimentation, and regional connectivity. By embracing innovation and sustainable technologies, the light aircraft segment of general aviation will continue to play a pivotal role in shaping the future of aviation, enabling safer, more efficient, and environmentally responsible flight operations. Their contribution to training, recreation, emergency services, and technological advancement ensures that general aviation remains a dynamic and essential component of the global aviation landscape.

4.2.1 Light Aircraft in General Aviation

Light aircraft are generally defined by their maximum takeoff weight (MTOW), typically under 12,500 pounds (5,670 kilograms) in the United States, although regulations vary by country. These aircraft often have one or two engines and can accommodate between two to six occupants, including the pilot. They are widely used for pilot training, personal transportation, aerial photography, agricultural applications, and emergency medical services.

Considerations in the Design of Light Aircraft

The appeal of light aircraft lies in their affordability, flexibility, and operational efficiency. Compared to larger aircraft, they require shorter runways, can access smaller airports, and have lower operational costs. Technological advancements in avionics, materials, and propulsion systems have further enhanced their performance, safety, and reliability. Light aircraft also serve as a critical entry point into aviation for aspiring pilots, providing foundational flight experience before transitioning to larger or commercial aircraft.

4.2.2 Design Characteristics of Light Aircraft

The design of light aircraft prioritizes simplicity, efficiency, and safety. Airframe construction often uses aluminum alloys or composite materials to balance strength and weight. Aerodynamic design emphasizes low drag and high lift-to-drag ratios, ensuring adequate performance at lower power levels. Fixed landing gear is common in smaller models to reduce complexity, though retractable gear may be employed in higher-performance variants to reduce drag. Engines are typically piston-powered, although turboprop and increasingly electric propulsion systems are being introduced. Piston engines are valued for their reliability, ease of maintenance, and lower fuel costs. Propeller selection and optimization are crucial for efficiency, especially for aircraft designed for training or recreational use. Additionally, modern light aircraft integrate glass cockpits and digital avionics systems, improving situational awareness, navigation, and safety.

4.2.3 Operational Roles of Light Aircraft

Light aircraft serve a diverse range of operational roles. Flight training is one of the most significant uses, with training aircraft providing the foundational skills for new pilots. These aircraft are often designed to be forgiving, easy to handle, and equipped with dual controls for instructor oversight. Recreational flying is another major segment of general aviation. Light aircraft allow private pilots to travel for leisure, participate in airshows, or enjoy aerial sightseeing. Additionally, aerial work operations such as crop dusting, aerial photography, surveying, and banner towing rely on light aircraft for flexibility and low operating costs. Emergency services, including air ambulance and disaster response, also use light aircraft to provide rapid transportation in areas with limited infrastructure.

4.2.4 Safety Considerations in Light Aircraft

Safety is paramount in general aviation, and light aircraft are no exception. While smaller aircraft are generally less complex, they are more susceptible to weather, pilot error, and mechanical issues. Regulatory bodies require regular inspections, maintenance schedules, and pilot certifications to ensure safe operation. Modern light aircraft are increasingly equipped with advanced avionics, such as GPS navigation, autopilot, traffic awareness systems, and collision avoidance technologies. These systems enhance situational awareness and reduce the likelihood of accidents, particularly in congested airspace or low-visibility conditions. Pilot training programs also emphasize risk management, decision-making, and adherence to standard operating procedures.

Table: Common Light Aircraft Types and Features

Aircraft Type	Typical Engine	Seating Capacity	Primary Use
Single-Engine Piston	Reciprocating	2–4	Training, recreation
Twin-Engine Piston	Reciprocating	4–6	Travel, small cargo
Turboprop	Turbine	4–6	Regional transport, emergency services
Light Sport Aircraft (LSA)	Piston	1–2	Recreational, sport flying
Electric Aircraft	Electric Motor	1–4	Training, experimental, short-range travel

4.2.5 Economic Impact of Light Aircraft

Light aircraft contribute significantly to the aviation economy. They support local airports, flight schools, maintenance operations, and aircraft manufacturing. Many regions rely on light aircraft for connectivity, especially in remote or underserved areas.

The affordability and lower operating costs of light aircraft make them accessible to small businesses, flying clubs, and individual owners, promoting broader participation in aviation. The industry also drives innovation, with research in fuel efficiency, electric propulsion, composite materials, and avionics often originating in the light aircraft segment before scaling to larger aircraft. This creates a feedback loop where general aviation supports technological advancement across the broader aerospace sector.

4.2.6 Environmental Considerations

Despite their smaller size, light aircraft contribute to greenhouse gas emissions and noise pollution. However, emerging electric and hybrid-electric propulsion systems are being developed to mitigate these environmental impacts. Electric light aircraft offer near-zero emissions for short flights, while quieter operation reduces noise impact on communities near airports. Sustainable aviation fuels (SAFs) are also being introduced for piston and turboprop engines, offering incremental reductions in carbon emissions. Airframe design improvements, such as laminar flow wings and lightweight composite materials, further enhance fuel efficiency. Flight planning and air traffic management optimizations also contribute to reduced environmental impact, particularly for training and recreational flights that may involve frequent takeoffs and landings.

4.2.7 Regulatory Framework for Light Aircraft

Light aircraft operations are governed by regulations set by national and international aviation authorities. In the United States, the Federal Aviation

Administration (FAA) oversees certification, pilot licensing, maintenance standards, and operational rules. Aircraft are certified under Part 23 regulations, which cover normal, utility, and commuter categories.

In Europe, the European Union Aviation Safety Agency (EASA) provides similar oversight. Regulations encompass airworthiness, pilot certification, flight operations, and safety management systems. These regulatory frameworks ensure that light aircraft meet safety and performance standards while providing flexibility for experimental or innovative designs, such as electric or light sport aircraft.

4.2.8 Technological Innovations in Light Aircraft

Technological advances are transforming the light aircraft segment. Electric propulsion is gaining traction, particularly for training aircraft and short-range recreational flights. Electric motors offer high efficiency, low noise, and reduced maintenance compared to piston engines.

Hybrid systems, combining batteries with small combustion engines, extend range and operational flexibility. Avionics advancements, including glass cockpits, GPS navigation, autopilot, and traffic awareness systems, enhance safety and reduce pilot workload. Composite materials improve strength-to-weight ratios, enabling lighter airframes and higher efficiency. Innovations in aerodynamics, such as winglets and laminar flow designs, further reduce drag and fuel consumption.

4.2.9 Training and Pilot Development

Light aircraft are the foundation of pilot training. Flight schools utilize small aircraft to teach basic flying skills, navigation, instrument procedures, and emergency management. Training aircraft are designed to be forgiving, with predictable handling characteristics and dual controls for instructors. Simulation technology complements in-flight training, allowing students to practice emergency procedures, instrument flying, and navigation in a controlled environment. Modern flight training programs emphasize safety culture, decision-making, and adherence to standard operating procedures, producing pilots prepared for both general aviation and commercial operations.

4.2.10 Light Aircraft in General Aviation

- **Affordability:** Lower purchase and operating costs compared to larger aircraft.
- **Operational Flexibility:** Capable of using small airports and short runways.
- **Diverse Applications:** Includes training, recreation, aerial work, and emergency services.

- **Simplified Maintenance:** Fewer systems and moving parts compared to commercial aircraft.
- **Technological Innovation:** Platform for testing electric propulsion, composites, and avionics.
- **Pilot Training Platform:** Provides foundational experience for new pilots.
- **Environmental Potential:** Electric and hybrid options reduce emissions and noise.
- **Regulatory Compliance:** Governed by national and international aviation authorities.
- **Safety Enhancements:** Modern avionics and safety systems improve operational safety.
- **Economic Contribution:** Supports airports, flight schools, maintenance, and aerospace manufacturing.

4.2.11 Challenges Facing Light Aircraft Operations

Despite their benefits, light aircraft face challenges. Fuel costs and availability can be limiting, particularly in remote regions. Weather sensitivity is higher than for larger aircraft, as smaller planes are more affected by turbulence and crosswinds. Pilot skill and experience are critical for safe operation, making training and proficiency essential. Infrastructure limitations, such as small airport capacity and lack of hangar space, can restrict operations. Regulatory compliance and insurance requirements may also pose barriers for recreational pilots or small operators. Additionally, transitioning to electric or hybrid propulsion introduces challenges related to energy density, battery weight, and charging infrastructure.

4.3 URBAN AIR MOBILITY (UAM) AND EVTOLS

4.3.1 Vertical Takeoff and Landing Technologies

Fundamental VTOL Configurations and Designs

The VTOL landscape encompasses diverse aircraft configurations, each with unique mechanisms for achieving vertical and forward flight. Tilt-rotor configurations represent one prominent design approach, featuring nacelles that house propellers and motors which tilt vertically for hover and horizontally for forward flight. This configuration offers significant advantages in cruise efficiency and speed, making it optimal for longer-range missions, though it requires complex mechanical tilting mechanisms and sophisticated control systems during the transition phase between flight modes. Tilt-wing designs take a different approach by tilting the entire wing along with its mounted propellers between vertical and horizontal positions. This configuration can provide more uniform airflow in hover compared to some designs but faces challenges with crosswind susceptibility and increased drag when the wing is vertical.

Other innovative configurations include tail-sitter aircraft that rest and take off on their tails, transitioning to level flight by pitching the entire fuselage forward. This design offers mechanical simplicity with no redundant actuators but presents challenges in control during transition and structural demands of landing on its tail. Ducted fan configurations feature a single rotor enclosed within a cylindrical duct, often combined with thrust vectoring for control. This approach offers enhanced safety and compact design with potential for high speed, though it can be susceptible to wind and requires sophisticated control laws. Lift-plus-cruise configurations represent another significant category, featuring dedicated vertical lift rotors for VTOL and a separate forward-facing propeller for cruise flight. This architecture provides simpler control as propulsion systems have dedicated functions, though the vertical lift rotors become dead weight during cruise, reducing overall efficiency.

Multifaceted Benefits of VTOL Technology

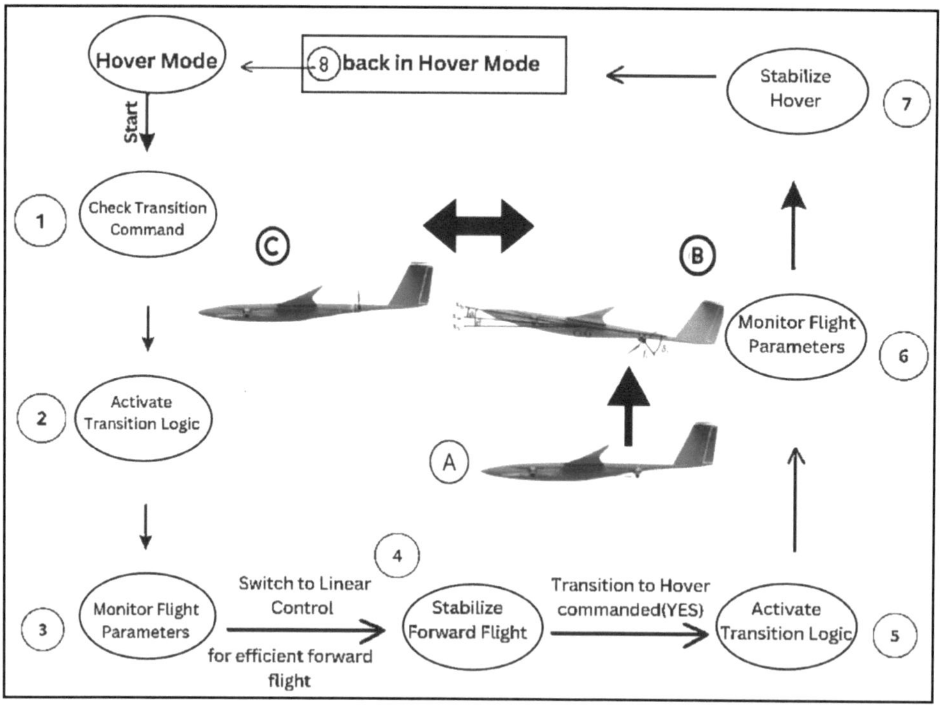

VTOL UAVs

The advantages of VTOL systems extend far beyond the simple ability to avoid runways, offering transformative benefits across multiple domains. The operational versatility and basing independence provided by VTOL capability allows aircraft to operate from confined or unprepared spaces, including rooftops in dense urban environments, ships at sea, or disaster zones where traditional infrastructure is damaged or non-existent.

This basing independence represents a critical enabler for new operational concepts across civilian and military applications. Hybrid VTOL unmanned aerial vehicles combine the hovering and vertical takeoff abilities of a multirotor with the speed and range of a fixed-wing aircraft, allowing a single platform to perform complex missions requiring both persistent station-keeping and high-speed area coverage. This makes them invaluable for applications including mapping, surveillance, and agricultural monitoring.

Enhanced safety through redundancy represents another significant advantage, particularly in electric VTOL designs that utilize distributed electric propulsion with multiple independent motors and rotors. This architecture provides inherent redundancy where, in the event of a single motor failure, the aircraft can often maintain controlled flight and land safely using its remaining motors, representing a significant safety improvement over single-engine helicopter. Noise reduction and community integration benefits arise from electric propulsion being inherently quieter than internal combustion engines used in most helicopters, while the ability to use multiple smaller, optimized rotors further helps in reducing noise signatures. This lower acoustic footprint is essential for gaining regulatory approval and public acceptance for frequent operations over populated areas.

- **Operational Versatility**: Enables operations from rooftops, ships, and disaster zones without requiring traditional runways
- **Mission Flexibility**: Combines hover capability with efficient forward flight for diverse mission profiles
- **Enhanced Safety**: Distributed propulsion provides redundancy for continued operation after component failures
- **Reduced Noise**: Electric propulsion and optimized rotor designs minimize acoustic impact on communities
- **Time Efficiency**: Bypasses ground congestion for significantly reduced urban transit times
- **Reduced Infrastructure Requirements**: Eliminates need for expensive and space-consuming runways
- **Environmental Benefits**: Electric variants produce zero operational emissions
- **Rapid Deployment**: Enables quick launch and recovery in emergency and military scenarios

- **Precision Operations:** Allows exact positioning for specialized applications like construction and surveying
- **Cost Effectiveness:** Reduced infrastructure needs and simpler maintenance for certain configurations

Electric Revolution: eVTOLs and Urban Air Mobility

The convergence of VTOL with electric propulsion has catalyzed the development of electric VTOL aircraft and the emerging vision for Urban Air Mobility. Electric motors offer several key advantages for VTOL applications, including higher reliability, simpler mechanical design, and zero operational emissions. The eVTOL sector has seen rapid growth with numerous companies developing aircraft designed to carry passengers and cargo in urban environments.

These aircraft typically feature innovative configurations optimized for both hover efficiency and cruise performance, with many utilizing distributed electric propulsion systems that enhance both safety and operational capabilities. The vision for Urban Air Mobility involves networks of these quiet, efficient eVTOLs operating from vertiports throughout cities, providing a fast, automated, and sustainable alternative to ground transportation. Urban Air Mobility represents a fundamental reimagining of urban transportation, promising to reduce travel times significantly for everyday journeys while alleviating ground congestion.

The development of this new transportation ecosystem requires advances not only in aircraft technology but also in ground infrastructure, air traffic management, and regulatory frameworks. Current eVTOL designs focus on achieving the necessary balance between payload capacity, range, and noise levels to make urban operations practical and socially acceptable. The progression from demonstration vehicles to certified commercial aircraft represents a significant engineering challenge that requires addressing complex issues of reliability, safety, and operational efficiency. As the technology matures, eVTOLs are expected to initially serve niche markets before expanding to broader transportation applications, ultimately transforming how people and goods move within and between urban centers.

Technical Challenges and Innovation Frontiers

Despite the exciting potential, widespread adoption of advanced VTOL technology faces several significant technical challenges that are the focus of intense research and development. The single greatest challenge for electric VTOL aircraft is the limited energy density of current battery technology. While conventional aviation fuels possess a vastly superior energy-to-weight ratio, even advanced lithium batteries store only a fraction of that energy, directly limiting the payload, range, and endurance of eVTOLs. Breakthroughs in battery chemistry are considered crucial for expanding the operational envelope of these aircraft, with research targeting significant improvements in energy density while maintaining safety and cycle life. The development of next-generation battery technologies represents a critical innovation frontier that will determine the ultimate capabilities and applications of electric VTOL systems.

Flight control and transition management present another major technical challenge, particularly during the phase where a VTOL aircraft transitions from vertical hover to horizontal forward flight. This transition represents an aerodynamically complex and control-intensive process that requires sophisticated and highly reliable flight control algorithms to manage the shifting dynamics and control authorities seamlessly and safely under all conditions. The development of these control systems involves addressing nonlinear aerodynamics, cross-coupling effects, and potential instability during the transition corridor. Noise mitigation remains a persistent challenge, as community acceptance depends on minimizing acoustic impact even though eVTOLs are generally quieter than helicopters. Research continues into optimizing rotor design, operational procedures, and flight paths to further reduce noise levels. Thermal management represents another critical area, as high-power electric motors and batteries generate significant heat during operation, especially during power-intensive takeoff and landing phases. Effective and lightweight thermal management systems are essential to maintain performance, ensure safety, and prolong the lifespan of critical components.

Diverse Applications across Industries

The unique capabilities of VTOL technologies are being leveraged across a wide spectrum of civilian and defense sectors, demonstrating the versatility and transformative potential of these systems. Urban Air Mobility and passenger transport represent one of the most prominent application areas, with development of piloted and autonomous passenger-grade eVTOLs for air taxi and ridesharing services forming the core of the urban air mobility vision. These applications aim to decongest city roads by moving a portion of transportation demand into the airspace above urban areas. The logistics and cargo delivery sector represents another major application area, where VTOL unmanned aerial vehicles are ideal for delivering medical supplies, parcels, and other goods to hard-to-reach locations. The potential for rapid urban parcel delivery continues to drive investment and innovation in this sector, pending regulatory approval for expanded operations.

Defense and security applications have emerged as early adopters of VTOL technology, with military forces worldwide adopting VTOL unmanned aerial vehicles for intelligence, surveillance, and reconnaissance missions. The ability to operate from confined spaces makes these systems perfect for border patrol, tactical reconnaissance, and force protection missions. Specialized military VTOL systems are designed with low visual, thermal, and acoustic signatures for covert operations, providing strategic advantages in sensitive environments. Precision agriculture and environmental monitoring represent another significant application area, where farmers use VTOL drones for crop health monitoring, pesticide application, and field mapping. The endurance and versatility of these systems allows them to cover large agricultural areas efficiently, revolutionizing farm management practices and enabling more sustainable agricultural methods. Emergency response and disaster management benefit tremendously from VTOL capabilities, as drones can be rapidly deployed in the immediate aftermath of natural disasters like floods, earthquakes, or wildfires to assess damage, locate survivors, and deliver critical aid where ground access is impossible and infrastructure is destroyed.

4.4 DRONES AND UNMANNED ELECTRIC AIRCRAFT

Drones and unmanned electric aircraft are moving from novelty to necessity, stitching together a new layer of airborne services. The winners will combine efficient airframes, disciplined energy management, resilient comms, and certifiable safety practices with operations that scale cleanly. Regulation is converging on risk-based frameworks that reward demonstrated reliability and transparency; UTM/U-space will turn fragmented corridors into predictable highways. Battery energy will improve steadily, but the near-term gains will come from smarter integration—larger, slower rotors, better thermal design, high-voltage architectures, and software that treats energy, airspace, and safety as a single optimization. Done well, unmanned electric aviation offers a rare trifecta: faster service, lower environmental impact, and quieter skies.

4.4.1 Landscape of Missions and Market Segments

Small multirotors dominate close-range tasks: visual inspections of bridges and towers, first-response overwatch, roof surveys, and creative cinematography. They excel at station-keeping and precision positioning, trading endurance for control authority and ease of use. In agriculture, multirotors and fixed-wing drones map fields with multispectral cameras, guiding fertilizer and irrigation with centimeter-level data. For logistics, lightweight winged VTOLs and fixed wings fly samples, pharmaceuticals, and consumer packages over 10–50 km routes, privileging reliability and airspace integration over raw payload.

At the higher end, larger unmanned aircraft systems (UAS) with hybrid-electric powerplants pursue hours-long inspections of pipelines and transmission corridors, maritime patrol, and regional cargo. High-altitude, long-endurance (HALE) solar-electrics promise persistent communications and sensing platforms measured in days, not hours. Each niche favors a different balance of airframe, energy storage, and autonomy, but they are united by the advantages of electric drive: instant torque, throttle precision, low vibration, and simple maintenance.

4.4.2 Airframes and Propulsion: Matching Layout to Mission

Airframes cluster into three families. Multirotors use multiple independently controlled rotors for thrust and control, with no wings. They hover efficiently but pay an endurance penalty in forward flight due to constant induced power. Fixed-wing drones trade hover for endurance, lifting their weight with aerodynamic surfaces; they cruise far on modest power but need runways, catapults, or nets. Vertical takeoff and landing (VTOL) hybrids combine both: tiltrotors tilt entire propulsion units to transition from hover to wing-borne flight; lift-plus-cruise designs dedicate rotors for vertical lift and separate propellers for forward thrust.

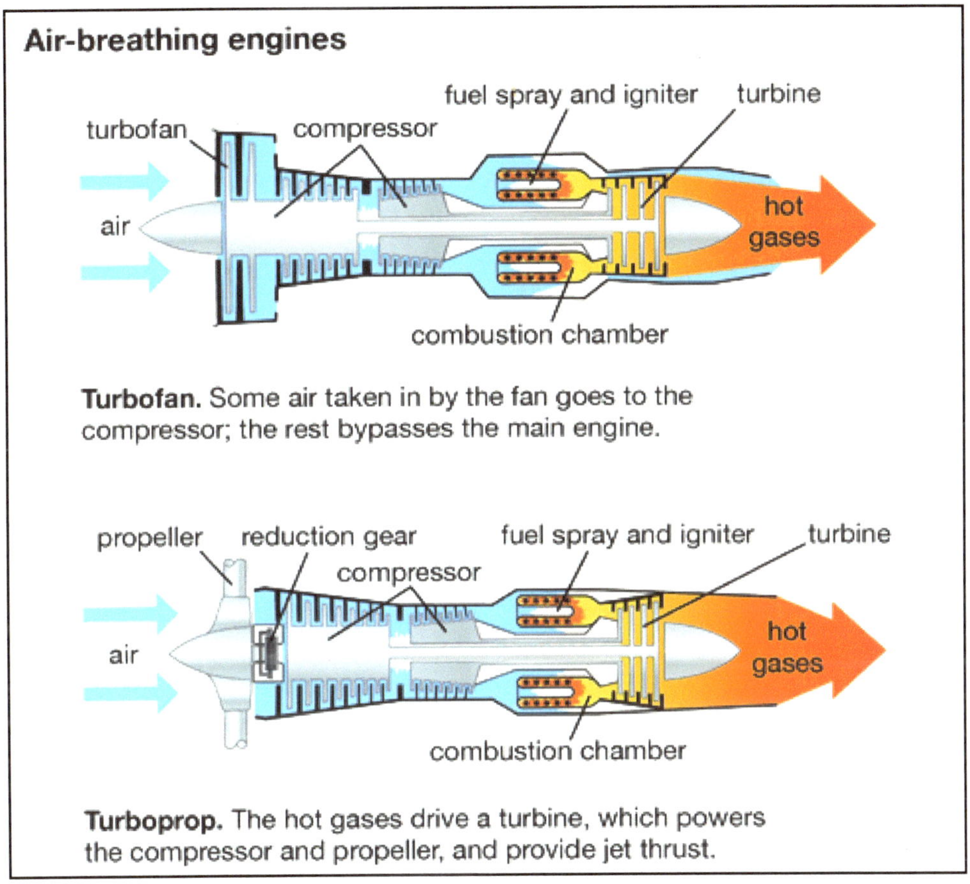

Turbofan. Some air taken in by the fan goes to the compressor; the rest bypasses the main engine.

Turboprop. The hot gases drive a turbine, which powers the compressor and propeller, and provide jet thrust.

Structural and Propulsion Aspects of Aircraft

Propulsor choices matter. Larger-diameter rotors at lower tip speeds reduce disk loading and hover power, improving both endurance and noise. Direct-drive motors eliminate gearbox losses and whine; when gears are necessary, helical or harmonic drives balance torque density with acoustic and efficiency goals. Distributed electric propulsion—multiple smaller propulsors—allows fine control of lift and redundancy, but adds weight and wiring that must be offset through careful integration and a high-voltage architecture to keep currents manageable.

4.4.3 Energy Systems and Battery Management

For most unmanned electric aircraft, lithium-ion chemistries (often high-nickel NMC) deliver the best mix of energy, power, and weight. Lithium iron phosphate (LFP) offers superior cycle life and safety at lower energy density, making it attractive for training fleets and high-cycle inspection operations. Pack design is as important as chemistry: high voltage lowers current and cable mass; modular packs enable quick swaps; robust thermal paths and liquid jackets reduce derates in hot climates and restore power in cold.

Effective energy is less than nameplate. Operators reserve a safety margin to handle diversions, winds, or hover-heavy recoveries, and they restrict state-of-charge windows to extend life. High C-rate events—vertical climbs, gusty hover, heavy payloads—cause voltage sag and heat, shrinking usable capacity unless managed with conservative power limits and preconditioned packs. Fleet software that tracks state-of-health (SOH), depth-of-discharge histograms, and temperature exposure enables predictive replacement and smarter dispatch, sending fresher packs on longer legs and protecting aging ones from abuse.

4.4.4 Autonomy, Navigation, and Flight Control

Autonomy stacks fuse perception, localization, planning, and control. GNSS with RTK/PPK yields centimeter-level positioning for mapping, while visual-inertial odometry and LiDAR maintain navigation under canopy or urban canyon conditions. Geofencing enforces airspace boundaries; terrain and obstacle databases seed safe routing; and onboard perception identifies dynamic hazards—cranes, birds, other drones—at short range. For dense tasks like facade inspection, planners must respect sensor geometry as well as flight constraints.

Controllers span from classical PID to model predictive control (MPC) and nonlinear techniques that handle rotor aerodynamics, wind gusts, and tiltrotor transitions. Health monitoring watches for motor imbalance, sensor drift, and battery anomalies, triggering graceful degradation modes: speed reductions, altitude holds, or autonomous return-to-home. For swarms, consensus and formation control add mesh-level robustness, but real-world operations still require strict separation and reliable deconfliction with crewed traffic.

4.4.5 Communications, C_2 Links, and BVLOS

Reliable command-and-control (C_2) is the spine of beyond-visual-line-of-sight (BVLOS) operations. Line-of-sight radios in unlicensed ISM bands work for short ranges but are vulnerable to interference. LTE/5G links extend reach and support high data rates for telemetry and video, leveraging network slicing and QoS to prioritize safety traffic where available. Satellite links (Lband and emerging lowEarthorbit services) provide redundancy in rural or maritime areas at the expense of latency and cost. BVLOS approval typically demands dual C_2 paths, link monitoring with rapid failsafe, and autonomous behavior when links degrade. Detect-and-avoid (DAA) solutions vary: ADSB In for cooperative traffic awareness, passive RF and transponder sniffers, radar or EO/IR for noncooperative detection, and network-based UTM feeds that broker strategic deconfliction. Robust cybersecurity—encrypted links, authenticated updates, secure boot—protects fleets from spoofing and hijacking risks that scale with remote operations.

4.4.6 Safety Engineering and Reliability

Electric systems invite redundancy by design. Multirotor drones can survive a motor failure if weight, control authority, and power reserves allow; that requires propeller-out flight testing and sufficient margins in ESC and bus design. Critical sensors—IMUs, barometers, magnetometers—are often duplicated and voted; dual GNSS receivers resist multipath and jams. Flight controllers trend toward safety-certifiable processes and hardware for larger UAS, with watchdogs and partitioned software that isolates non-critical functions. Mechanical and electrical safety extend to the ground. Fire-safe battery handling, charging enclosures, and pack quench protocols are essential given lithium thermal risks. Parachute recovery systems, tested to standards like ASTM F3322, can reduce ground risk and unlock operations over people. Maintenance regimes track vibration signatures, bearing temperatures, insulation resistance, and connector wear, shifting from reactive repair to condition-based maintenance that preserves dispatch reliability.

4.4.7 Regulation and Airspace Integration

Rules differ by region and aircraft class, but patterns are clear. In the United States, Part 107 governs small UAS under 55 lb for line-of-sight operations, with waivers for night, over-people, and BVLOS; Remote ID is now required for most operations. Commercial delivery carriers have leveraged Part 135 to conduct package delivery with higher assurance, under specific operating specifications. Europe's EASA framework spans Open, Specific, and Certified categories, with the Specific Operations Risk Assessment (SORA) defining mitigations for medium-risk missions and U-space services enabling controlled integration in designated airspace. Larger unmanned aircraft entering controlled airspace move into certified categories approaching crewed aircraft rigor.

Across jurisdictions, Remote ID, geofencing, and UTM/U-space integrations are becoming table stakes. Approval hinges on demonstrated operational safety—procedures, training, maintenance—and evidence that detect-and-avoid, C_2 robustness, and contingency handling meet the risk profile. Early engagement with regulators and standardized documentation accelerates approvals, as does using platforms already vetted for similar missions.

4.4.8 Infrastructure, Fleet Ops, and Scaling

Scaling from pilots to production means treating drones as a fleet. "Drone docks" and nests automate launch, landing, and charging or battery swaps, enabling 24/7 coverage with minimal human touch. Standardized charge connectors, weatherized pads, and robotic pack handling shrink turnaround variance. Operations centers monitor multiple aircraft, ingesting telemetry, weather, NOTAMs, and UTM directives, with alerting that prioritizes exceptions and hands control to human operators when needed. Data pipelines matter as much as flight logs. Images and sensor data require secure, bandwidth-efficient upload and labeling; model retraining loops improve perception over time. Spare parts, propeller balancing, ESC calibration, and firmware management demand logistics discipline familiar to aviation but sized for drones' shorter cycles and lower unit cost. The payoff is high utilization—more missions per aircraft per day—at consistent quality and safety.

4.4.9 Environmental Performance and Noise

Unmanned electric aircraft reduce point-of-use emissions and cut noise relative to combustion engines, a key benefit for urban and suburban operations. The acoustic signature is dominated by rotor/propeller noise, which designers mitigate with larger, slower rotors, refined blade shapes, and RPM scheduling that avoids tonal stacking. Flight procedures that minimize hover and avoid hard throttle changes reduce annoyance and energy consumption simultaneously.

Lifecycle impacts include battery materials and manufacturing; diligent pack management extends life, and second-life stationary storage plus high-recovery recycling reduce footprint. On a perpackage or perinspection basis, electric drones often beat vans or helicopters on emissions and energy use—especially when charged with lowcarbon power—while also avoiding road congestion and improving response times for urgent deliveries.

4.4.10 Economics and Business Models

- **Electric Propulsion Lowers Variable Costs:** electricity is inexpensive relative to fuel, motors are durable, and maintenance is simple. The big levers are battery depreciation, utilization, and labor. High cycle counts with shallow depth-of-discharge, moderate charge rates, and good thermal control extend battery life; automated docks and remote operations lift utilization per operator. For logistics, cost per delivery hinges on route density, payload fraction, airspace access, and service level agreements; drones shine where time value is high (medical) or where ground routes are slow or costly.
- **Partnerships Shape Unit Economics:** Utilities and railroads internalize inspection benefits; hospitals and pharmacies value time-certain

medical logistics; retailers' co-site micro-fulfillment with drone pads to compress last-mile costs. Standards-based platforms reduce integration time; battery-as-a-service decouples capex; and data services—maps, asset health, inventory updates—create recurring revenue beyond flight fees.

4.4.11 Design and Operations Levers for Electric Drones

- Right-size the airframe to the mission; avoid oversizing that drags endurance down with unused payload and motor capacity.
- Use low disk loading with larger, slower rotors to reduce hover power and noise; tip Mach and blade loading are first-order drivers.
- Raise bus voltage to cut current and harness mass; fewer I^2R losses extend endurance and improve thermal margins.
- Instrument batteries deeply and precondition before flight; precise SOE estimates enable smaller reserves without compromising safety.
- Plan energy-aware routes that minimize hover and climb, exploit tailwinds, and use continuous descent profiles to land with healthy margins.
- Build dual, dissimilar C_2 links and automatic failsafes; autonomous return-to-home and safe-landing site selection protect BVLOS missions.
- Incorporate propulsor-out capability and test it; verify ESC, inverter, and thermal headroom to survive a motor failure without loss of control.
- Standardize charging and swapping at "drone docks"; automate turnaround to increase utilization while controlling battery stress.
- Align maintenance with telemetry-driven triggers—vibration, temperatures, error counts—instead of fixed hours; it cuts surprises and cost.
- Engage communities with transparent noise and route data; adjust procedures and schedules to maintain social license as operations scale.

4.5 MILITARY AND EXPERIMENTAL ELECTRIC AIRCRAFT

4.5.1 Historical Development and Early Experiments

Military experimentation with electric aircraft traces back several decades, though early efforts were severely limited by battery technology and power system efficiency. Initial programs focused primarily on small unmanned aerial vehicles where endurance requirements were modest and payload capacity minimal. These early electric UAVs demonstrated the potential for silent operation and reduced infrared signatures, characteristics particularly valuable for covert surveillance missions. The Defense Advanced Research Projects Agency (DARPA) and similar organizations worldwide recognized these advantages early, funding research programs that laid groundwork for today's more ambitious projects.

The evolution from simple battery-powered demonstrators to sophisticated experimental platforms reflects dramatic improvements in enabling technologies. Energy storage density has increased manifold, while electric motors have achieved power-to-weight ratios approaching those of turbine engines in certain applications. Control systems leveraging artificial intelligence and advanced materials enable aircraft configurations impossible with conventional propulsion. These technological advances transformed electric aircraft from curiosities to serious contenders for specific military missions, prompting established aerospace companies and innovative startups alike to pursue development programs with increasing urgency and resources.

4.5.2 Current Military Electric Aircraft Programs

Army Electric Aircraft Research takes Flight in Partnership

Contemporary military electric aircraft programs span a remarkable range of sizes and missions. Small tactical UAVs utilizing electric propulsion have become commonplace, offering company and battalion-level units organic surveillance capabilities with minimal logistics burden. These systems operate from austere locations, require minimal maintenance, and provide hours of nearly silent overwatch capability. Larger experimental platforms explore the boundaries of electric flight, including high-altitude pseudo-satellites designed to operate at stratospheric altitudes for months at a time, serving as persistent surveillance platforms or communications relays. Several nations have initiated programs to develop electric training aircraft, recognizing the potential for dramatic reductions in pilot training costs. Electric trainers offer consistent performance characteristics, reduced maintenance complexity, and the ability to operate from airfields with strict noise restrictions. The simplified systems in electric aircraft also provide excellent platforms for initial flight training, allowing students to focus on fundamental flying skills without the complexity of engine management. Advanced programs explore hybrid-electric configurations for larger military transports and special operations aircraft, seeking to combine the efficiency of electric propulsion with the energy density of conventional fuels for extended range missions.

4.5.3 Technical Challenges and Solutions

The primary technical challenge facing military electric aircraft remains energy storage density. Current lithium-ion battery technology, while vastly improved from earlier generations, still falls short of the energy density provided by aviation fuel by approximately two orders of magnitude. This fundamental limitation drives innovative approaches to aircraft design and mission planning. Engineers pursue multiple strategies to mitigate this challenge, including advanced aerodynamics to minimize power requirements, lightweight structures using cutting-edge composite materials, and operational concepts that leverage the unique advantages of electric propulsion while avoiding its limitations. Power management and thermal control present additional significant challenges.

High-power military applications generate substantial heat that must be dissipated without compromising aircraft performance or signature management. Advanced cooling systems, including phase-change materials and integrated thermal management architectures, address these challenges while adding minimal weight. Safety considerations for high-voltage systems in combat environments necessitate robust protection schemes and redundancy levels exceeding civilian standards. Electromagnetic compatibility requirements in military platforms, where sensitive electronic warfare and communications systems must coexist with high-power electric drives, demand careful system integration and shielding strategies.

4.5.4 Innovative Design Approaches

Electric propulsion enables radical departures from conventional aircraft configurations. Distributed electric propulsion, where multiple small motors replace single large engines, offers unprecedented flexibility in aircraft design. This approach enables improved aerodynamic efficiency through propulsion-airframe integration, enhanced redundancy for combat survivability, and novel control mechanisms that blur the distinction between propulsion and flight control systems. Experimental military programs explore configurations with dozens of electric motors embedded in wing structures or distributed around airframes in ways impossible with conventional engines.

Morphing aircraft structures represent another area where electric systems enable revolutionary capabilities. Electric actuators can alter wing geometry, control surface effectiveness, or even overall aircraft configuration in flight, optimizing performance for different mission phases. Variable-pitch propellers driven by electric motors can reverse thrust instantly, enabling extremely short landing distances and unique maneuvering capabilities. Some experimental programs investigate biomimetic designs inspired by natural flyers, using arrays of small electric actuators to achieve wing movements resembling those of birds or insects, potentially enabling unprecedented agility and efficiency.

4.5.5 Power Systems and Energy Storage

Beyond conventional batteries, military experimental programs investigate diverse energy storage and generation technologies. Fuel cells offer higher energy density than batteries while maintaining the silent operation and low thermal signature advantages of electric propulsion. Hybrid systems combining batteries for high-power demands with fuel cells for cruise flight extend operational capabilities while maintaining many electric propulsion advantages. Solar cells integrated into wing structures supplement battery power for high-altitude, long-endurance missions, with some experimental aircraft achieving perpetual flight capability under favorable conditions.

Supercapacitors provide rapid power delivery for high-performance maneuvers or emergency situations, complementing battery systems in military applications requiring brief bursts of extreme performance. Advanced battery chemistries, including solid-state designs and lithium-metal configurations, promise substantial improvements in energy density and safety. Military research programs also investigate exotic energy storage methods, including mechanical systems like high-speed flywheels and compressed air storage, seeking solutions optimized for specific mission profiles where conventional batteries fall short.

4.5.6 Testing and Evaluation Programs

Military experimental electric aircraft undergo rigorous testing regimes that push systems to extremes rarely encountered in civilian applications. Environmental testing subjects aircraft to temperature extremes, vibration, shock, and electromagnetic interference levels simulating combat conditions. Flight testing explores the entire performance envelope, including edge cases where electric systems might behave unexpectedly. Long-duration tests validate reliability and maintenance predictions critical for operational deployment decisions. Operational evaluation programs place experimental electric aircraft in realistic military scenarios, assessing their effectiveness against current capabilities and doctrine. These evaluations often reveal unexpected advantages or limitations not apparent in laboratory testing. For instance, the nearly silent operation of electric aircraft has proven more tactically significant than initially anticipated, enabling new concepts of operation for special forces insertion and extraction missions. Conversely, limitations in cold-weather battery performance have prompted development of specialized heating systems and operational procedures for arctic deployments.

4.5.7 Strategic and Tactical Implications

The adoption of electric propulsion in military aviation carries profound implications for force structure and operational concepts. Reduced logistics footprints enable expeditionary forces to operate from austere locations with minimal support infrastructure. The simplicity of electric powertrains reduces maintenance personnel requirements and spare parts inventories, critical factors in deployed operations. Silent operation enables new tactical approaches to air assault, reconnaissance, and close air support missions where acoustic signature previously limited operational options.

Strategic implications extend to industrial base considerations and technology security. Nations leading in electric aviation technology gain potential military advantages that may shift regional power balances. The dual-use nature of many electric propulsion technologies complicates export control regimes and technology transfer policies. Dependencies on rare materials for batteries and motors create new supply chain vulnerabilities requiring strategic planning and stockpiling. These considerations drive national investment strategies and international collaboration frameworks as nations seek to maintain technological superiority while managing proliferation risks.

CHAPTER 5

Infrastructure and Charging Ecosystem

5.1 GROUND CHARGING TECHNOLOGIES AND STANDARDS

Ground charging technologies and standards form the backbone of electric aviation's infrastructure. They ensure that electric aircraft can operate efficiently, safely, and sustainably across diverse flight profiles and airport environments. From DC fast chargers to wireless inductive systems, advancements in power delivery and interoperability are enabling faster turnarounds and broader adoption of electric flight. The development of universal standards through global cooperation ensures compatibility and safety, while smart grid integration and renewable energy adoption make charging operations more sustainable. Though challenges persist—particularly in scalability and infrastructure investment—the trajectory is clear: the future of aviation will rely heavily on intelligent, high-efficiency ground charging systems. Combining innovation in technology, safety certification, and standardization, electric aviation is paving the way toward a cleaner, quieter, and more connected future. Ground charging technologies are not merely logistical tools—they are enablers of a new era of sustainable air transportation that aligns environmental stewardship with operational excellence.

5.1.1 Ground Charging in Electric Aviation

Electric aviation relies on stored electrical energy, typically in high-capacity lithium-based batteries, to power propulsion systems. Once aircraft complete their flight cycles, they must be recharged quickly to ensure operational efficiency, particularly for commercial or regional operations where turnaround time directly affects profitability. Ground charging technologies serve this purpose by providing controlled power delivery from the electrical grid or renewable sources to the aircraft's onboard energy storage system.

The challenges are significant. Aircraft require much higher energy densities and charging rates than electric vehicles (EVs). For example, a small electric trainer aircraft may require 50–100 kWh per flight, while regional electric aircraft could demand several megawatt-hours. Ground charging systems must therefore balance speed, safety, reliability, and standardization, while integrating with existing airport infrastructure and energy systems.

5.1.2 Evolution of Ground Charging Technologies

The evolution of ground charging technologies in aviation follows patterns similar to the electric automotive industry but with unique aviation-specific adaptations. Early prototypes used modified electric vehicle chargers, suitable for low-capacity aircraft. As the energy requirements of electric aircraft increased, manufacturers began developing aerospace-grade charging systems with higher power outputs, better insulation, and improved safety mechanisms.

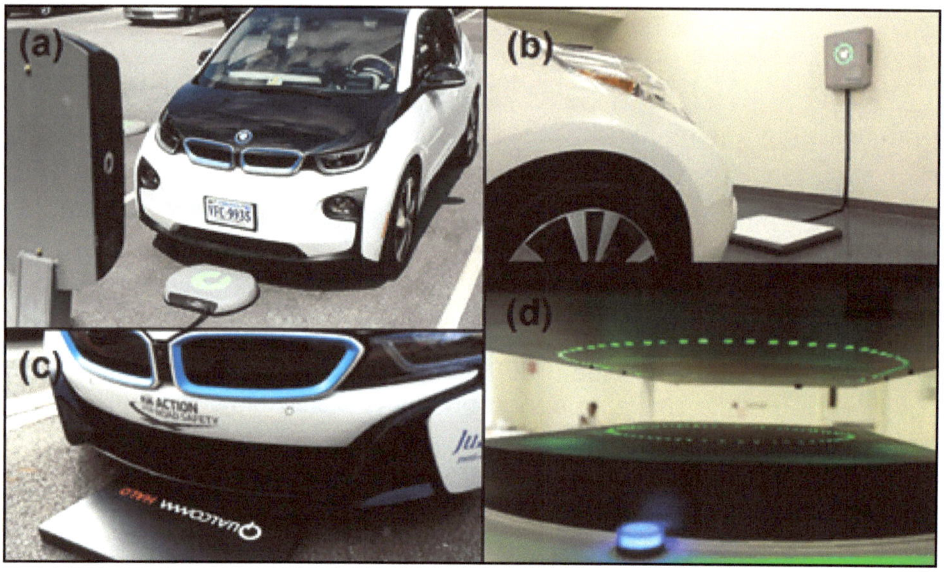

Trends and Emerging Technologies for the Development of Electric Vehicles

Recent developments include high-voltage direct current (DC) fast chargers capable of delivering hundreds of kilowatts to several megawatts. Some experimental systems use modular charging stations, allowing simultaneous charging of multiple aircraft or scaling power delivery according to demand. Wireless and inductive charging methods are also under exploration, promising automated, contactless power transfer suitable for urban air mobility (UAM) and drone operations.

5.1.3 Charging Interface and Connector Standards

A key aspect of ground charging technology is the charging interface—the physical and electrical connection between the charging station and the aircraft. Standardization of connectors and communication protocols is critical to ensure compatibility across different aircraft manufacturers and airport systems. Efforts are underway by international organizations such as SAE International, EUROCAE, and ICAO to develop standardized interfaces. The SAE AE-7D committee, for instance, is working on guidelines for electric aircraft charging standards similar to the SAE J1772 used in electric cars. These standards cover voltage levels, connector pin configurations, safety interlocks, and communication requirements between the charger and aircraft. Standardized connectors not only facilitate interoperability but also improve safety by preventing mismatched voltage levels or incompatible plug types. They enable the development of universal charging stations, reducing infrastructure costs and promoting widespread adoption.

Table: Comparison of Ground Charging Technologies

Charging Method	Power Range	Efficiency	Applications
AC Charging	20–100 kW	Moderate	Light aircraft, training aircraft
DC Fast Charging	100–1000 kW	High	Regional aircraft, rapid turnaround
Megawatt Charging	1–5 MW	Very High	Large electric aircraft, commercial use
Wireless (Inductive) Charging	10–300 kW	Moderate	UAM, drones, automated systems
Battery Swap Systems	Variable	High	Rapid operations, minimal downtime

5.1.4 Power Delivery and Grid Integration

Electric aircraft charging places significant demands on the electrical grid. A single high-capacity charger can draw as much power as several homes or even a small industrial facility. Therefore, grid integration and management are essential. Airports are developing smart grid systems capable of dynamically managing energy flow between the grid, renewable sources (like solar arrays), and storage systems. These systems can balance loads, reduce peak demand charges, and ensure that charging operations do not overwhelm local electrical infrastructure. Some airports also employ microgrids—localized energy systems combining solar panels, battery storage, and backup generators—to provide stable, renewable power for charging stations. Energy management software helps optimize charging schedules based on flight timetables, energy prices, and grid conditions. This coordination ensures efficient energy usage while maintaining high aircraft availability.

5.1.5 Safety Protocols and Certification

Safety is paramount in aviation, and ground charging systems must adhere to strict certification and safety standards. Charging operations involve high voltages—often exceeding 1,000 volts—and large currents, which present electrical, thermal, and fire risks if not properly managed. International safety standards address several aspects, including insulation, grounding, fault detection, and emergency shutdown mechanisms. Charging systems incorporate temperature monitoring, current limitation, and real-time diagnostics to prevent overheating or overcharging. Connectors are designed with mechanical interlocks to prevent disconnection during active charging, and automated shutoff systems ensure safe operation even in adverse weather conditions. Regulatory authorities such as the Federal Aviation Administration (FAA) and European Union Aviation Safety Agency (EASA) are also working with manufacturers to develop certification pathways for charging infrastructure, ensuring compliance with electrical and aviation safety requirements.

5.1.6 Aspects of Ground Charging Technologies

- **High Power Output:** Systems must deliver hundreds of kilowatts to megawatts for efficient turnaround.
- **Standardized Connectors:** Ensures interoperability across manufacturers and airports.
- **Smart Grid Integration:** Balances energy use between charging systems and airport operations.
- **Safety and Certification:** High-voltage systems must meet aviation-grade safety standards.
- **Scalability:** Infrastructure should accommodate future increases in aircraft capacity.
- **Energy Storage Systems:** On-site batteries help buffer demand and stabilize power delivery.
- **Renewable Integration:** Solar and wind energy can support sustainable charging operations.
- **Fast and Efficient Charging:** Reduces turnaround time and improves operational viability.
- **Automated and Wireless Options:** Enables hands-free charging for UAM and autonomous aircraft.
- **Data Communication Protocols:** Supports smart diagnostics, billing, and load management.

These features collectively define the efficiency, safety, and sustainability of ground charging systems in electric aviation.

5.1.7 Battery Swapping as an Alternative

While conventional plug-in charging remains the most common method, battery swapping offers an alternative for rapid turnaround operations. Instead of recharging onboard batteries, aircraft can exchange depleted battery packs for fully charged ones in minutes. This system eliminates waiting time for recharging and can be automated using robotic handling systems. However, battery swapping requires standardized battery modules and storage facilities at airports. It also demands precise mechanical design to ensure quick, safe exchanges. While it is promising for small aircraft and drones, scalability for larger aircraft remains a challenge due to the size and weight of battery packs. Nonetheless, as battery technologies evolve toward modularity and lightweight materials, swapping could become a viable solution for time-sensitive operations.

5.1.8 Wireless and Inductive Charging Technologies

Wireless or inductive charging is another innovation being explored for electric aviation. It uses electromagnetic fields to transfer energy between a ground-based charging pad and a receiver coil on the aircraft. This contactless method minimizes wear and tear on connectors, eliminates human error during plug-in, and can enable automated charging for drones, air taxis, and short-haul aircraft. While current power transfer efficiencies are slightly lower than direct plug-in systems, advances in resonant inductive coupling and high-frequency inverters are improving performance. Wireless charging is especially suited for urban air mobility (UAM) environments, where frequent short flights and fast automated turnarounds are essential.

5.1.9 Global Standards Development

Establishing international standards is crucial to ensure compatibility and scalability. Organizations like SAE International (Society of Automotive Engineers), EUROCAE (European Organisation for Civil Aviation Equipment), and ICAO (International Civil Aviation Organization) are leading efforts to harmonize charging standards.

The SAE AE-7D Aerospace Committee is developing standards for high-voltage DC charging systems, focusing on connectors, communication protocols, and safety interlocks. EUROCAE's WG-112 addresses similar topics, ensuring alignment with European regulations. Collaboration between these organizations and regional authorities ensures that charging systems developed in one part of the world can be safely and efficiently used elsewhere.

5.1.10 Infrastructure Implementation at Airports

Implementing ground charging infrastructure at airports involves strategic planning, investment, and integration with existing systems. Airports must consider the location of charging stations, energy demand management, and safety zones for high-voltage operations. Some airports are introducing dedicated charging zones near taxiways or hangars for electric aircraft fleets. Airports are also exploring renewable microgrids, combining solar arrays with on-site storage to provide sustainable charging power. Such systems not only reduce carbon emissions but also enhance energy independence and resilience. For regional airports, modular charging units that can be easily relocated or scaled up offer flexibility for future growth.

5.2 BATTERY SWAPPING AND MODULAR SYSTEMS

Battery swapping and modular systems represent a fundamental reimagining of how energy is stored, managed, and replenished in electric aviation. By addressing the critical challenge of aircraft downtime through rapid energy exchange rather than slow recharging, these technologies enable operational patterns that were previously impractical for electric aircraft. The modular approach further enhances this capability by introducing unprecedented levels of flexibility, maintainability, and scalability to aircraft energy systems.

While significant challenges remain in standardization, infrastructure investment, and certification, the compelling operational and economic benefits are driving continued innovation and adoption across multiple aviation segments. The urban air mobility sector particularly stands to benefit from these technologies, as the business case for high-frequency urban air taxi services depends critically on minimizing ground time between flights. As the industry matures, battery swapping and modular systems are likely to evolve from specialized solutions to mainstream approaches, ultimately transforming not only how electric aircraft are operated but also how the supporting infrastructure is designed and deployed. The successful implementation of these technologies will require close collaboration among aircraft manufacturers, battery producers, infrastructure developers, and regulatory authorities, but the potential rewards in terms of operational efficiency, economic viability, and environmental sustainability make this collaboration eminently worthwhile. As electric aviation continues its rapid development, battery swapping and modular systems will play an increasingly central role in realizing the full potential of emission-free flight.

5.2.1 Technical Foundations of Battery Swapping Systems

Battery swapping technology operates on the principle of mechanical exchange rather than electrical recharging, fundamentally changing how energy is replenished in electric aircraft. The core infrastructure consists of specialized swapping stations equipped with automated machinery designed to safely handle heavy battery packs with precision and speed. These stations incorporate sophisticated alignment systems that guide the aircraft into exact positioning for the swap procedure, followed by robotic mechanisms that disconnect and remove the depleted battery before installing a fresh unit. The process is managed by integrated control systems that verify electrical connections, monitor battery health parameters, and ensure proper mechanical engagement throughout the operation. Safety systems include multiple redundancy features, thermal monitoring, and fail-safe mechanisms to prevent accidental disconnection or improper installation. The swapping interface itself requires careful engineering to accommodate high-power electrical connections, cooling system linkages, and secure physical mounting points that can withstand flight loads. Modern swapping systems utilize standardized connectors and mounting configurations to ensure interoperability across aircraft types, though achieving true universal standards remains an industry challenge. The entire swap process is designed to be completed within timeframes comparable to conventional refueling, typically targeting five to ten minutes from aircraft arrival to departure with a fresh energy source.

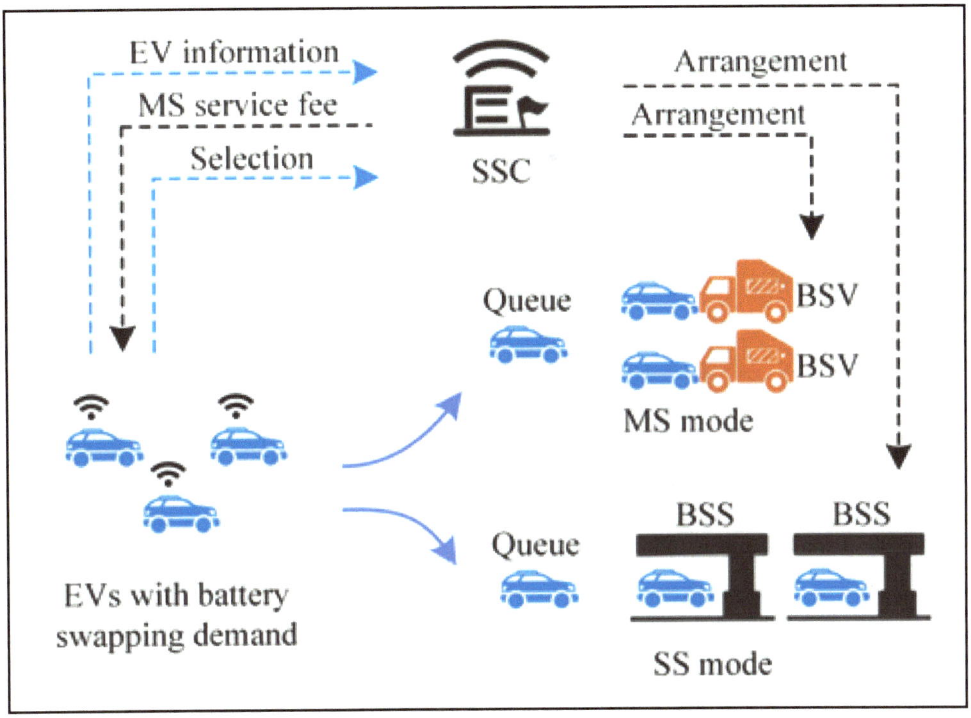

Engineering Principles of Battery Swap Systems

5.2.2 Architectural Principles of Modular Battery Systems

Modular battery systems represent a fundamental redesign of how energy storage is conceptualized and implemented in electric aircraft. Rather than utilizing a single, monolithic battery pack, these systems employ multiple smaller, self-contained battery modules that work together as an integrated power source. Each module contains its own battery cells, battery management system, thermal controls, and structural housing, creating a functionally independent unit that can operate as part of a larger system. The modular architecture allows for flexible configuration to meet different aircraft power requirements, with modules being combined in series and parallel arrangements to achieve the necessary voltage and capacity. This approach provides significant advantages in manufacturing, as standardized modules can be produced at scale for use across multiple aircraft platforms. Maintenance benefits are equally substantial, as individual faulty modules can be identified, removed, and replaced without taking the entire battery system out of service. From a safety perspective, modular designs incorporate physical and electrical isolation between modules, helping to contain any thermal events and prevent cascading failures. The modular approach also enables graceful degradation, where the failure of a single module reduces overall capacity but doesn't necessarily render the entire system inoperable. This architectural philosophy extends to the swapping infrastructure itself, where modular designs facilitate handling, storage, and charging of standardized energy units.

5.2.3 Operational Benefits and Economic Advantages

The implementation of battery swapping and modular systems delivers substantial operational and economic benefits that directly address key challenges in electric aviation. The most immediate advantage is the dramatic reduction in aircraft downtime, as swapping a battery takes minutes compared to the hours required for charging large battery packs. This enables electric aircraft to achieve significantly higher utilization rates, making them economically competitive with conventional aircraft for high-frequency operations.

The modular approach further enhances operational flexibility by allowing operators to tailor battery capacity to specific mission requirements, installing more modules for longer routes and fewer for shorter hops. From a financial perspective, these systems enable new business models such as Battery-as-a-Service, where operators pay for energy usage rather than bearing the full capital cost of battery ownership. This model transfers the burden of battery degradation and technology obsolescence to specialized service providers, reducing financial risk for aircraft operators.

Maintenance efficiency is greatly improved through modular designs that enable targeted replacement of faulty components rather than requiring full system overhaul. The distributed nature of modular systems also provides operational redundancy, as the failure of individual modules typically results in reduced performance rather than complete system failure. For aircraft manufacturers, modular architectures simplify certification processes by allowing standardized testing of individual modules across multiple aircraft platforms. The combination of these benefits creates a compelling economic case for adopting swapping and modular approaches, particularly in applications where aircraft utilization and operational flexibility are critical to business viability.

- **Minimized Aircraft Downtime:** Battery swapping reduces ground time from hours to minutes, enabling electric aircraft to achieve utilization rates comparable to conventional aircraft.
- **Reduced Capital Expenditure:** Battery-as-a-Service models decouple energy storage costs from aircraft acquisition, lowering barriers to entry for operators.

- **Extended Battery Lifespan:** Centralized charging and maintenance at swapping stations optimize battery health and prolong usable service life.
- **Enhanced Operational Safety:** Modular designs contain failures and prevent cascading thermal events, while swapping stations provide controlled environments for battery handling.
- **Improved Maintenance Efficiency:** Faulty modules can be individually identified and replaced without taking entire systems out of service.
- **Increased Operational Flexibility:** Battery capacity can be tailored to specific missions, and swapping stations provide rapid energy replenishment across networks.
- **Future-Proof Technology Integration:** Modular systems allow incremental adoption of improved battery technologies as they become available.
- **Optimized Second-Life Applications:** Standardized modules facilitate repurposing of aviation batteries for less demanding stationary storage applications.
- **Streamlined Infrastructure Deployment:** Modular swapping stations can be deployed more rapidly than extensive charging networks.
- **Superior Fleet Utilization:** Rapid turnaround enables higher daily flight frequencies, improving revenue potential for commercial operators.

5.2.4 Implementation Challenges and Technical Hurdles

Despite the compelling advantages, widespread implementation of battery swapping and modular systems faces significant technical and operational challenges that must be addressed. Standardization represents perhaps the most formidable obstacle, as the industry lacks universal specifications for battery dimensions, electrical interfaces, and mounting systems. Without common standards, swapping infrastructure would need to support multiple incompatible battery types, undermining the economic efficiency of the approach.

The development of swapping stations requires substantial capital investment in automated handling equipment, charging infrastructure, and inventory management systems. These stations must operate with extremely high reliability to support continuous aircraft operations, necessitating robust designs with redundant systems and comprehensive maintenance protocols. From a technical perspective, the repeated mechanical engagement and disengagement of high-power electrical connections presents engineering challenges related to contact wear, thermal management, and connection integrity. The weight penalty associated with modular designs, including additional housing, connectors, and management systems, must be carefully managed to avoid unacceptable impacts on aircraft performance. Certification authorities require rigorous demonstration of safety and reliability for both the modular battery systems and the swapping processes, creating additional development timelines and costs.

Thermal management during the swapping process requires careful design, as batteries may need to be handled shortly after high-power discharge operations. The logistics of managing battery inventory, tracking state of health, and coordinating charging cycles present complex operational challenges that require sophisticated software systems and predictive analytics. Finally, the development of business models that appropriately allocate costs and risks among aircraft manufacturers, battery providers, infrastructure operators, and aircraft operators remains an ongoing challenge that must be resolved for these systems to achieve widespread adoption.

5.2.5 Industry Applications and Use Cases

Battery swapping and modular systems find particularly compelling applications in specific segments of the electric aviation market where operational intensity and flexibility requirements align with the strengths of these technologies. Urban air mobility represents perhaps the most natural application, with electric vertical takeoff and landing aircraft operating high-frequency routes in dense urban environments. In this context, minimal ground time is essential for economic viability, making rapid battery swapping a critical enabling technology. Regional air connectivity presents another promising application, where smaller electric aircraft serving short-haul routes can benefit from quick turnarounds at multiple airports without requiring extensive charging infrastructure at each location.

Unmanned cargo delivery operations represent an early adoption market, where automated drones can implement battery swapping to enable continuous operation with minimal human intervention. Flight training organizations constitute another ideal use case, as training aircraft typically operate multiple short-duration flights per day with brief intervals between sessions. In these applications, the ability to quickly swap batteries between lessons enables much higher aircraft utilization than would be possible with conventional charging. Military and special operations represent additional promising applications, where modular power systems enable mission-specific configuration and rapid turnaround in forward operating environments. In each of these use cases, the specific implementation of swapping and modular systems may vary, but the core benefits of reduced downtime, operational flexibility, and economic efficiency remain consistent drivers for adoption. As these technologies mature, they are likely to expand into additional applications where the operational patterns align with the strengths of modular, swappable energy systems.

5.3 AIRPORT INFRASTRUCTURE FOR ELECTRIC AVIATION

Airport infrastructure for electric aviation is a multidimensional challenge—electrical, thermal, digital, operational, and social. The airports that move fastest will approach it as a phased, standards-driven program: start with right-sized power and modular chargers, add storage to tame peaks, integrate thermal preconditioning and predictive scheduling, and wrap it all in strong safety and cybersecurity practices. As aircraft capabilities improve and standards converge, those foundations will support higher turn power, more pads and gates, and deeper integration with airline and vertiport networks. The result is not just a new plug on the ramp, but a resilient energy ecosystem that unlocks quieter operations, lower emissions, and reliable, scalable electric flight.

5.3.1 Grid Connection and Capacity Planning

The first step is understanding load. A single regional electric aircraft might charge at several hundred kilowatts; a bank of eVTOL pads turning every few minutes can push instantaneous demand into the multi-megawatt range. Airports need load forecasts that reflect peaky profiles—morning and evening waves—and variability from weather, utilization, and route mix. Capacity planning works backward from the turn-time targets: how many simultaneous charges at what power, what buffer is needed for delays, and how resilient must the system be under equipment faults or grid events. Upgrades often include dedicated medium-voltage feeders, new substations, and ring-fed distribution for redundancy. Early utility engagement is critical—substation siting, interconnection studies, and tariff design can take years, and co-investment models vary by jurisdiction. Airports that plan in phases—provisioning conduits, transformer pads, and protection gear ahead of load—avoid costly retrofits as fleets scale.

5.3.2 Megawatt-Class Charging and Power Quality

Electric aircraft rely on high-voltage, high-power DC charging to minimize turn time. Power quality matters: harmonics, voltage sag, and flicker can interfere with sensitive airport systems and tenant operations. Modern rectifiers with active front ends and power factor correction help meet interconnection requirements, while local filters and staged charging strategies smooth peaks. Airports should segment charging into zones with independent feeders and protection, limiting fault propagation and enabling maintenance without taking the entire system offline. For vertiports, compact power cabinets near pads shorten cable runs and reduce I^2R losses; for regional gates, centralized DC plants feeding dispensers via DC busways can simplify maintenance. As charging power rises, liquid-cooled cables or pantograph-style connectors become necessary for ergonomics and heat management. Future-proofing means routing for higher voltage and anticipating megawatt-class connectors even if initial operations start with lower power.

5.3.3 Energy Storage, Microgrids, and On-Site Generation

Grid capacity is not the only lever. Co-located energy storage systems (ESS) buffer peaks, reduce demand charges, and provide ride-through during grid disturbances. Batteries sized to a few minutes of peak pad operation can shave the tallest spikes; larger systems enable off-peak charging and even limited islanded operation for continuity during outages.

Microgrid control orchestrates ESS, solar PV on roofs and carports, and, where appropriate, dispatchable generation such as fuel cells or low-NOx generators. Airports can monetize flexibility through demand response and capacity markets while guaranteeing charging availability for priority flights. Safety and siting for ESS require careful attention—clearances, ventilation, monitoring, and firefighting access—especially in airfield environments with strict security and operational constraints.

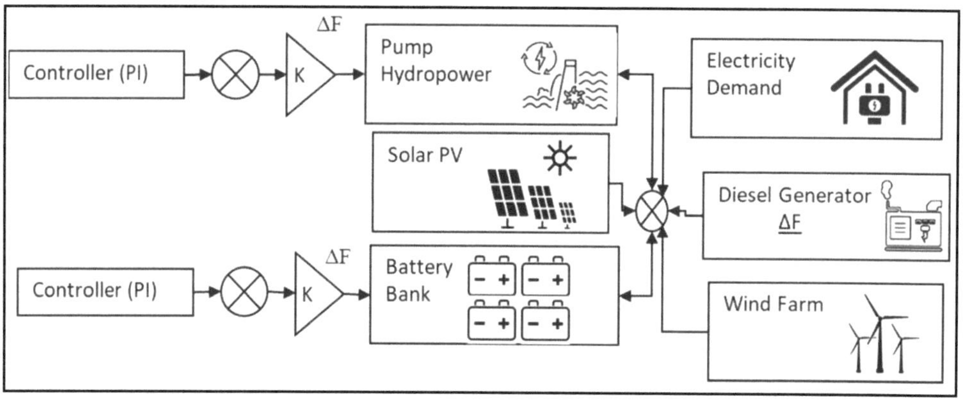

Battery Storage, Microgrid Systems, and On-Site Power Supply

5.3.4 Thermal Management and Preconditioning

Heat is the hidden currency of electric operations. Batteries, inverters, and motors all have thermal envelopes that constrain power and longevity. Airports can reclaim usable energy and reduce in-flight derates by preconditioning packs at the stand. Chilled or tempered glycol loops within the charger plant, coupled to aircraft-side heat exchangers, bring cells into the optimal temperature window before departure. In hot climates, centralized chillers with thermal storage (ice tanks or phase-change materials) can shift cooling loads off-peak while enabling high power during turn. In cold climates, hangar preheat and charger-integrated warmup prevent power loss and lithium plating risks. Thermal design must include heat rejection—dry coolers or evaporative towers sized for simultaneous peak—with noise abatement and plume management so that cooling infrastructure remains a good neighbor.

5.3.5 Safety, Fire Protection, and Codes

Lithium battery safety shapes infrastructure from the ground up. Chargers and ESS need gas detection, thermal cameras, and rapid isolation to manage thermal runaway risks; water supply and application strategies should prioritize sustained cooling rather than smothering, with runoff containment and environmental controls for contaminated water. Clear e-stop systems, lockout/tagout procedures, and signage for high-voltage areas are essential. Airports will adapt standards traditionally used for heliports and fuel farms to electric operations, layering in electrical codes, ESS guidance, and aviation-specific safety assessments. Ventilation, separation distances, and blast panels for ESS rooms, along with arc-flash studies and appropriate PPE for ground crews, create a defensible safety case. Regular drills with airport fire and rescue services ensure readiness and close the loop between design intent and operational reality.

5.3.6 Ground Operations, Layout, and Vertiport Design

Physical layout determines throughput and safety. eVTOL pads need adequate spacing for downwash, obstacle clearance, and safe egress paths; approach/departure corridors should avoid building recirculation zones that amplify noise and turbulence. Charging pedestals must be positioned to prevent cable trips, accommodate different door heights and nose-to-pad orientations, and keep clear of rotor arcs. For regional aircraft, gate electrification may displace conventional ground power and pre-conditioned air carts, simplifying the ramp. Clear separation between charging and passenger areas, combined with barriers or retractable cable management, reduces human–machine conflicts. Vertiport design should integrate acoustic treatments—berms, canopies, and non-reflective surfaces—and provide visual cues for pilots and ground staff as operations scale from one or two pads to a matrix of simultaneous turns.

5.3.7 Interoperability, Standards, and Connectors

Standardization is the antidote to stranded assets. Airports should favor charging hardware and communication protocols that are vendor-agnostic and support over-the-air updates. Interoperable high-voltage DC connectors, secure charger-aircraft handshakes, and common data models for energy metering enable mixed fleets and competitive energy pricing without bespoke engineering for each operator. For cargo and passenger throughput, automated or semi-automated connectors reduce turn time and variability, but they must negotiate securely and verify safe states before energizing. Emerging megawatt-class standards will evolve; infrastructure should be modular enough to swap dispensers and cables without uprooting upstream power and civil works. Harmonizing billing, authentication, and scheduling across tenants turns charging into a utility-like service rather than a bespoke accommodation.

5.3.8 Digital Systems: EMS/SCADA, Scheduling, Cybersecurity

Electric aviation introduces a data layer that rivals air traffic systems in importance. An airport energy management system (EMS) coordinates charger, ESS, chillers, PV, and tariffs in real time, while SCADA provides visibility and control with alarms and trends for operators. Integrations with airline or vertiport scheduling, aircraft energy management systems (AEMS), and weather feeds enable predictive charging that aligns power delivery with arrival times, pack health, and departure slots. Cybersecurity must be designed in: network segmentation, encrypted protocols, secure boot for chargers and controllers, certificate management, and incident response plans protect critical infrastructure from intrusion. Data governance matters too—metering accuracy for billing, anonymization for shared analytics, and availability SLAs for tenants relying on digital services to meet their schedules.

5.3.9 Business Models, Tariffs, and Financing

The economics of airport electrification hinge on demand charges, utilization, and capital structure. Airports can pursue third-party energy-as-a-service providers to fund and operate charging and microgrids, trading capex for long-term service fees. Tariff design with utilities should reflect the public-benefit nature of aviation electrification and the operational need for reliable high-power windows—time-of-use rates, coincident peak relief, and standby charges all affect per-turn cost. On-site renewables and ESS may qualify for incentives and tax credits that improve payback; carbon accounting and environmental attribute certificates create additional value streams for operators marketing low-carbon flights. Transparent, predictable pricing builds tenant confidence and encourages fleet commitments that, in turn, justify infrastructure phases.

5.3.10 Workforce, Training, and Procedures

People make infrastructure work. Ground crews need training in high-voltage safety, connector handling, thermal preconditioning workflows, and emergency response. Maintenance teams must develop competencies in power electronics, chillers, and ESS—skills adjacent to but distinct from conventional GSE. Clear SOPs for charge authorization, cable management, FOD sweeps around charging pedestals, and coordination with tower/operations prevent small issues from cascading into delays. As automation grows—robotic connectors, drone docks, and autonomous tugs—procedures must cover human–robot interaction, overrides, and safe recovery from faults. Investing early in training and certification programs builds a bench of talent that can scale with operations.

5.3.11 Environmental and Community Considerations

Electrification reduces local emissions and, with thoughtful design, overall noise. Airports can maximize benefits by pairing clean power sources with operations scheduled to minimize community impact. Noise modeling for vertiport corridors, selection of cooling equipment with low acoustic signatures, and landscaping that doubles as acoustic shielding all help maintain social license. Lifecycle considerations—battery logistics, recycling pathways, and ESS embodied carbon—should be part of procurement criteria. Public dashboards that show noise events, charging carbon intensity, and flight counts build trust and align airport sustainability reports with observable community outcomes.

5.3.12 Infrastructure Levers for Ready-to-Scale Airports

- Engage utilities early to co-plan substations, feeders, and tariffs; design ring-fed, N+1 distribution to keep operations running through faults.
- Deploy on-site energy storage to shave peaks, provide ride-through, and reduce demand charges; size ESS to pad waves, not just total daily energy.
- Standardize on modular, interoperable megawatt-class DC chargers with liquid-cooled cables and secure comms for mixed-fleet compatibility.
- Integrate an EMS with SCADA, AEMS, and schedules to forecast arrivals and precondition batteries, aligning power and thermal capacity with turns.
- Build thermal plants—chillers, dry coolers, or heat pumps—with buffer storage to handle simultaneous high-power turns in hot or cold conditions.
- Engineer safety-in-depth: gas and thermal detection, e-stops, runoff containment, and trained ARFF protocols for lithium incidents and ESS rooms.
- Design vertiports and gates for flow: pad spacing for downwash, cable management, barriered charge zones, and ergonomic connector reach.
- Use microgrids with PV and, where feasible, fuel cells or low-emission generators to enhance resilience and lower lifecycle carbon intensity.
- Adopt clear pricing and SLAs; consider energy-as-a-service to accelerate deployment and de-risk capex while guaranteeing uptime for tenants.
- Institutionalize training for HV operations, codify SOPs, and run regular drills; treat chargers and ESS like critical aeronautical systems, not simple outlets.

5.4 GRID INTEGRATION AND RENEWABLE ENERGY SOURCES

Grid integration is the unsung foundation of the global energy transition. For all the excitement around falling solar costs and offshore wind growth, none of these advances will matter if the grid cannot handle the variable, decentralized power they produce. The ten critical considerations outlined here—from forecasting accuracy to cross-regional synchronization—provide a roadmap for governments, utilities, and investors to build a grid that is flexible, resilient, and capable of supporting 80% renewable electricity by 2050.

This is not a trivial task: it will require overcoming NIMBY opposition to transmission lines, reforming outdated regulatory processes, and investing in technologies that are still in their early stages. But the alternative—stalled renewable growth and continued reliance on fossil fuels—is far costlier. As the IEA notes, every dollar invested in grid integration today yields $3 in long-term decarbonization benefits. The grid is no longer just a network for delivering power—it is the backbone of a sustainable future.

5.4.1 Renewables Surge and the Grid Integration Gap

To understand the scale of the grid integration challenge, consider the dissonance between renewable deployment and grid capacity. In California, for example, solar farms generate so much excess power during midday summer hours that the state's grid operator often pays neighboring states to take the surplus—curtailing up to 10% of solar output annually—because the grid lacks the infrastructure to store or transmit the energy to evening peak demand. In Germany, a similar issue plagues offshore wind: the country's northern coast has enough wind capacity to power 40 million homes, but outdated transmission lines to southern industrial hubs mean 15% of that wind power is wasted each year.

These inefficiencies stem from a core design flaw: traditional grids rely on "baseload" power sources that provide constant output, with peaker plants (natural gas) filling gaps in demand. Renewables reverse this dynamic: supply fluctuates, and the grid must adapt demand or store excess power to maintain stability. This requires more than just adding renewables to the grid—it requires reengineering the grid itself to be flexible, digital, and interconnected. The IEA estimates that global grid investment will need to double to $600 billion annually by 2030 to keep pace with renewable growth; without this investment, renewables will struggle to reach even 50% of global electricity supply by 2050.

5.4.2 Critical Grid Integration Considerations for Renewable Energy Deployment

Hyper-local intermittency forecasting: Unlike fossil fuel plants, renewable output is tied to weather patterns—even a 10-minute cloud cover event can cut solar output by 50%. Grid operators need AI-powered, hyper-local forecasts (down to 1-kilometer resolution) to predict these fluctuations; the IEA notes that improving forecast accuracy by just 5% can reduce renewable curtailment by 10% and cut reliance on peaker plants.

- **High-Voltage Direct Current (HVDC) Transmission Expansion:** Most large-scale renewables are located in remote areas—offshore wind farms, desert solar arrays—far from urban load centers. HVDC lines lose 3-5% of power over long distances (compared to 8-10% for alternating current lines), making them critical for moving renewable power efficiently. The EU's North Sea Wind Power Hub, a planned network of HVDC lines connecting 10 countries, will unlock 150 GW of offshore wind by 2050.
- **Grid-Scale Energy Storage Scaling:** To shift excess renewable power from midday (solar peak) to evening (demand peak), grid-scale storage capacity must grow 20x by 2030, per IEA targets. While lithium-ion batteries dominate current storage, emerging technologies like flow batteries (with 10+ hour discharge times) and pumped hydro (which accounts for 90% of existing storage) are critical for long-duration needs.
- **Distributed Energy Resource (DER) Coordination:** Rooftop solar, home batteries, and electric vehicle (EV) chargers are decentralized resources that can either strain the grid or support it—depending on management. Smart grid platforms that aggregate DERs into "virtual power plants" (VPPs) allow grid operators to draw on home batteries during peak demand, reducing the need for peaker plants.
- **Inertial and Frequency Regulation:** Fossil fuel turbines provide "inertia"—rotational energy that stabilizes grid frequency (50 or 60 Hz) during supply fluctuations. Renewables lack this inertia, so grid operators need fast-acting reserves (e.g., battery storage that can respond in milliseconds) to prevent blackouts. Australia's Hornsdale Power Reserve, a 150 MW battery, has cut grid frequency deviations by 90% since 2017.
- **Demand Response Program Optimization:** Paying consumers and businesses to reduce power use during peak demand (e.g., turning off industrial machinery or adjusting home AC settings) is a low-cost way to balance renewable supply gaps. Texas' ERCOT demand response program prevented widespread blackouts during the 2023 heatwave by reducing peak demand by 2 GW—equivalent to the output of two coal plants.

- **Interconnection Queue Reform**: In the U.S., more than 1,400 GW of renewable projects are stuck in interconnection queues—waiting for grid operators to assess their impact—with average wait times exceeding three years. Reforming queues to prioritize high-impact projects (e.g., offshore wind near load centers) and streamline technical reviews could cut wait times by 50%.
- **Cross-Regional Grid Synchronization**: Connecting regional grids allows areas with surplus renewable power to share with areas in shortfall. For example, synchronizing the U.S. East and West coast grids would let California draw on Texas wind power during winter evenings, when solar output is zero. The IEA estimates that cross-regional synchronization can increase renewable penetration by 15% without additional storage.
- **Grid Digitalization and Edge Computing**: Smart sensors, real-time data analytics, and edge computing enable grid operators to detect and resolve issues (e.g., voltage fluctuations) before they escalate. McKinsey estimates that grid digitalization can reduce operating costs by 15% and cut outage durations by 30%—critical for integrating variable renewables.
- **Cybersecurity Hardening**: Renewable assets and smart grids are more digitally connected than traditional power plants, making them vulnerable to cyberattacks. Grid operators must implement end-to-end encryption, redundant control systems, and regular penetration testing to prevent attacks that could disable renewable generation or disrupt grid stability.

5.4.3 Emerging Technologies Bridging the Integration Gap

While these considerations outline the challenge, emerging technologies are already proving their ability to overcome grid integration barriers. Virtual power plants (VPPs) are a prime example: in Australia, the Tesla Virtual Power Plant aggregates 30,000 home batteries into a single resource that can supply 250 MW of power during peak demand—enough to power 50,000 homes. In Europe, HVDC undersea cables are connecting offshore wind farms to multiple countries, reducing curtailment by allowing excess power to be routed to the highest-demand markets.

Another breakthrough is "grid-forming inverters"—devices that allow renewable generators to mimic the inertial support of fossil fuel turbines. Unlike traditional "grid-following" inverters, which rely on the grid for frequency reference, grid-forming inverters can stabilize the grid independently. The U.S. Department of

Energy is testing these inverters at utility-scale solar farms, with early results showing they can prevent blackouts during renewable supply fluctuations.

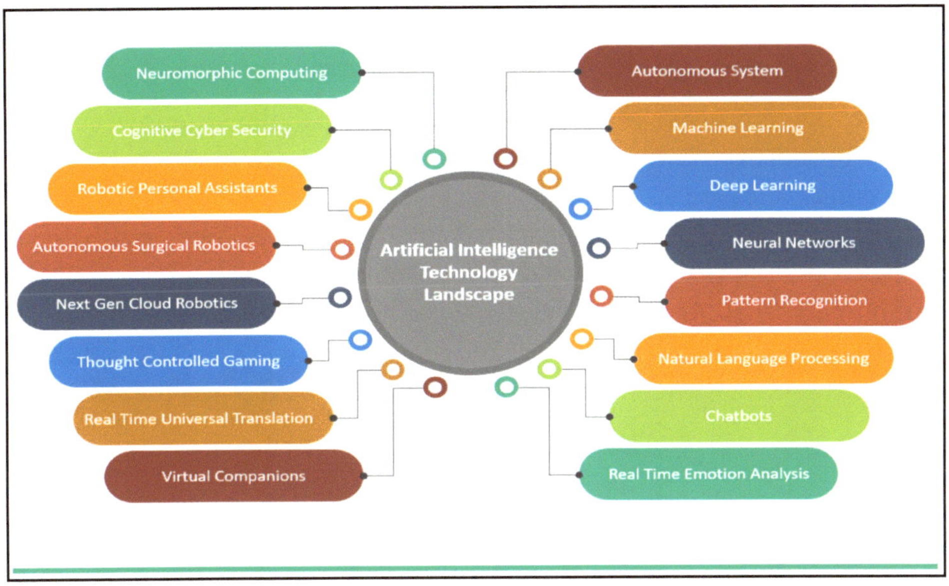

Advancing Technologies to Bridge Integration Challenges

5.4.4 Policy and Regulatory Enablers for Grid Integration

Technology alone cannot solve the grid integration challenge policy and regulation must create the framework for investment and innovation. In the U.S., FERC Order 2222, issued in 2020, requires grid operators to allow DERs to participate in wholesale energy markets, unlocking revenue streams for home battery owners and VPP operators.

The Inflation Reduction Act (2022) also allocates $10 billion for grid modernization, including funding for HVDC lines and grid-forming inverters. In the EU, the Clean Energy Package mandates that member states upgrade their grids to accommodate 42% renewable electricity by 2030, with penalties for failing to meet interconnection targets. These policies are critical: without clear regulatory signals, utilities and investors will hesitate to fund the $10 trillion in grid investment the IEA estimates is needed by 2050.

5.5 URBAN VERTIPORTS AND CHARGING HUBS FOR EVTOLS

Urban vertiports and charging hubs for eVTOLs stand at the intersection of aviation innovation, renewable energy, and smart city design. They are essential enablers of the electric aviation revolution, translating technological progress into tangible infrastructure. While challenges persist—ranging from regulatory harmonization to energy integration—the global push toward sustainable air mobility ensures that vertiports will become defining features of future urban landscapes. By merging advanced engineering with thoughtful planning, vertiports can make urban air travel safe, efficient, and environmentally responsible, heralding a new era of connected cities in the skies.

5.5.1 Concept and Design of Urban Vertiports

A vertiport is a compact aviation facility designed to handle the unique operational demands of eVTOLs. It typically consists of landing pads, passenger handling zones, maintenance areas, and most crucially, charging infrastructure for electric propulsion systems. The design must balance efficiency, safety, and integration within the dense fabric of urban landscapes. Unlike traditional airports, vertiports need minimal real estate and can be built on rooftops, near highways, or at existing transportation hubs. Their modularity and adaptability make them ideal for phased development in cities where space and regulatory flexibility are limited.

5.5.2 Infrastructure Components of Vertiports

The infrastructure of a vertiport is a blend of aviation safety principles and electric mobility technologies. Core components include landing and takeoff zones, navigation and lighting systems, charging stations, control centers, and passenger terminals. Safety protocols demand features such as wind and obstacle sensors, automated flight control guidance, and fire suppression systems designed specifically for lithium-ion or hydrogen-based energy sources. Charging stations, which can range from plug-in systems to wireless charging pads, form the technological heart of these facilities, ensuring operational continuity of eVTOL fleets throughout the day.

5.5.3 Integration with Urban Ecosystems

For vertiports to function effectively, they must be seamlessly integrated into existing transportation and energy ecosystems. Integration involves connecting vertiports with public transit, highways, and rail systems, allowing passengers to transition effortlessly between modes of travel. Urban planners and policymakers are working to ensure that vertiports align with sustainability goals and zoning regulations. Moreover, integration into the urban energy grid is essential, as charging hubs require substantial electrical capacity and resilience. Smart grid integration, supported by renewable energy sources, enhances both reliability and environmental performance.

5.5.4 Charging Hubs and Energy Infrastructure

Charging hubs within vertiports are designed to handle the high energy demands of rapid-turnaround eVTOL operations. Depending on aircraft design, charging may require anywhere from tens to hundreds of kilowatts per vehicle. Therefore, the establishment of robust, high-capacity electrical connections and battery management systems is fundamental. Charging hubs may utilize fast-charging technologies for rapid energy replenishment, swappable battery systems for efficiency, or wireless induction pads to reduce mechanical wear and improve automation. Future hubs are also exploring energy storage buffers, such as stationary batteries or supercapacitors, to balance load and reduce stress on urban power grids.

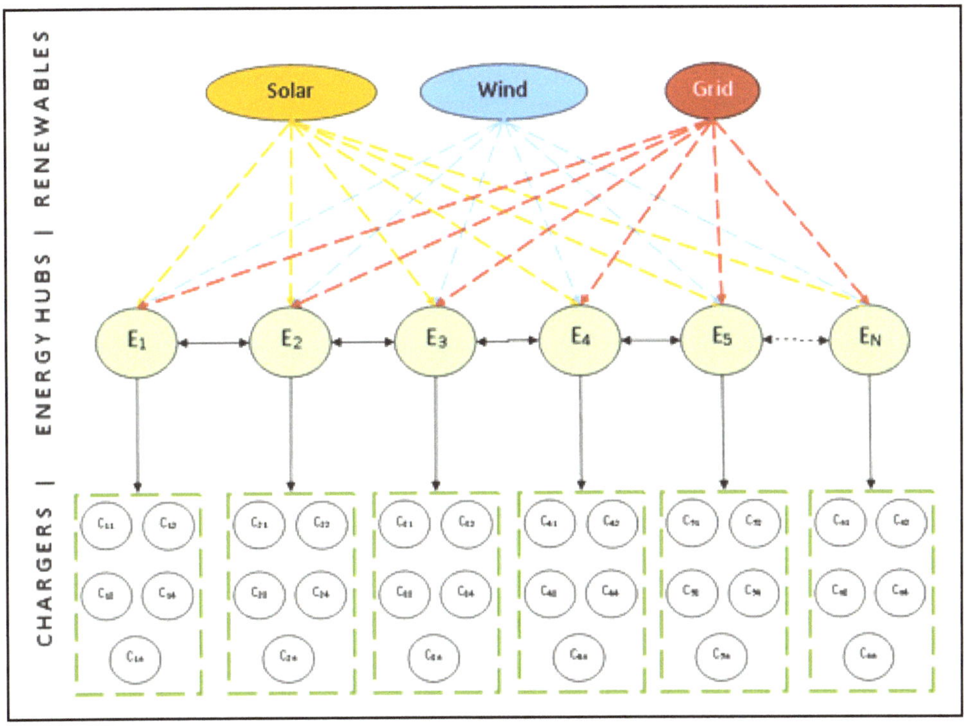

Utilizing Rooftop Renewable Energy Potential for Electric Vehicle Charging Infrastructure

Table: Vertiports vs. Traditional Heliports

Feature	Vertiport (for eVTOLs)	Heliport (for Helicopters)
Energy Source	Electric (battery/hybrid)	Fossil fuel (aviation fuel)
Noise Level	Low due to electric propulsion	High rotor noise
Infrastructure Footprint	Compact and modular	Larger, with fuel storage requirements
Environmental Impact	Zero local emissions	High carbon emissions
Integration Potential	Easily adaptable to urban settings	Restricted by noise and safety zones

5.5.5 Urban Planning and Site Selection

Selecting optimal locations for vertiports involves balancing accessibility, safety, and environmental impact. Key criteria include proximity to business districts, airports, and transportation corridors. Rooftop sites, waterfront platforms, and redeveloped industrial zones are often ideal due to their open airspace and reduced noise sensitivity. Urban planners must also consider local weather conditions, wind patterns, and emergency access routes. Simulation-based modeling helps identify potential air traffic conflicts and ensures compliance with aviation safety standards. Importantly, the success of vertiports hinges on community acceptance, requiring early public engagement to address concerns over safety, noise, and visual impact.

5.5.6 Regulatory and Safety Considerations

Developing urban vertiports demands an intricate interplay of aviation regulations, building codes, and energy standards. Authorities such as the Federal Aviation Administration (FAA), the European Union Aviation Safety Agency (EASA), and local urban authorities are drafting new frameworks for eVTOL infrastructure certification. Regulations cover aspects like landing pad dimensions, lighting requirements, fire protection, and charging station standards. Safety protocols must address potential battery hazards, electromagnetic interference, and emergency evacuation procedures. Additionally, cybersecurity for automated flight operations and charging systems is becoming an integral part of vertiport safety architecture.

5.5.7 Technological Innovations in Vertiport Development

The future of vertiport technology is evolving rapidly, driven by advancements in automation, artificial intelligence, and materials engineering. Automated guidance systems are being developed to manage multiple simultaneous takeoffs and landings within confined urban airspace. Digital twin models allow engineers to simulate operations, assess performance, and optimize vertiport layouts before construction. Furthermore, AI-driven energy management systems dynamically allocate charging capacity to minimize downtime and optimize grid usage. Innovations in lightweight modular structures and smart building materials enhance construction efficiency and sustainability, while autonomous ground handling robots streamline aircraft maintenance and passenger logistics.

5.5.8 Deployment and Standardization

Despite the promise of vertiports, several challenges hinder large-scale deployment. High infrastructure costs, inconsistent regulatory frameworks, and the need for standardization across eVTOL manufacturers complicate early adoption. Energy demands are also significant—cities must upgrade grid capacity to handle the additional load. Furthermore, operational economics remain uncertain, as return on investment depends heavily on passenger demand, regulatory stability, and public acceptance. Noise perception, although lower than helicopters, still poses community resistance in densely populated areas. Finally, interoperability among charging systems is critical; without universal standards, operators risk fragmentation and inefficiency in UAM networks.

5.5.9 Vertiport and Charging Hub Development

- Enables decentralized, point-to-point air mobility within cities
- Reduces ground congestion and travel time for commuters
- Provides clean, quiet alternatives to conventional air transport
- Facilitates integration with renewable energy and smart grids
- Supports scalable UAM ecosystems across metropolitan areas
- Creates new urban infrastructure and energy-related jobs
- Encourages innovation in autonomous flight and AI systems
- Improves emergency response and medical transport capabilities
- Promotes sustainable design and green infrastructure policies
- Strengthens collaboration between aviation and urban planning sectors

5.5.10 Economic and Environmental Impacts

The economic potential of vertiports extends far beyond aviation. They are poised to become multimodal transportation nodes that stimulate investment in smart city infrastructure, energy storage, and digital services. By enabling faster intra-city connections, vertiports can reshape business patterns and urban land use, fostering new economic clusters around mobility hubs. Environmentally, their all-electric operation reduces greenhouse gas emissions and air pollution, contributing to urban sustainability goals. However, ensuring that the electricity used is renewable remains a key factor in realizing these benefits. Lifecycle analysis of eVTOLs and vertiport infrastructure helps assess the full environmental footprint and guide policy decisions.

CHAPTER 6

Certification, Safety, and Regulation

6.1 GLOBAL REGULATORY BODIES (FAA, EASA, DGCA)

The FAA, EASA, and DGCA will continue to play central roles in shaping the future of global aviation safety as the industry evolves through technological transformation and operational changes. While each authority operates within its specific national or regional context, their collective influence establishes the safety foundation that enables international air transportation. The ongoing challenge for these organizations lies in balancing their fundamental safety mandates with the need to accommodate innovation, support industry efficiency, and address emerging priorities such as environmental sustainability.

The increasing complexity of aviation technologies and operations will require continued evolution in regulatory approaches, with greater emphasis on performance-based regulations, risk-based oversight, and international cooperation. The harmonization of regulatory standards will remain an important objective, though achieving complete uniformity appears unlikely given different national priorities, legal frameworks, and operational environments. As new entrants and technologies continue to emerge, the regulatory bodies will need to maintain their technical capabilities and develop new expertise to effectively oversee increasingly diverse aviation ecosystems. Despite these challenges, the fundamental mission of these organizations remains constant: to establish and maintain the highest practicable levels of safety in civil aviation through effective regulation, oversight, and international cooperation. Their continued adaptation and collaboration will be essential for ensuring that aviation safety keeps pace with industry transformation while maintaining public confidence in air transportation worldwide.

6.1.1 Federal Aviation Administration (FAA): America's Aviation Authority

The Federal Aviation Administration serves as the primary aviation regulatory body for the United States, governing one of the world's largest and most complex airspace systems. Established in 1958, the FAA has evolved into a comprehensive organization with responsibilities spanning aircraft certification, air traffic control, pilot licensing, and commercial operator oversight. The agency's regulatory framework is codified in Title 14 of the Code of Federal Regulations, which details requirements for every aspect of aviation operation from aircraft design to maintenance procedures and pilot training standards.

The FAA's certification processes are recognized globally, making them particularly significant for manufacturers and operators engaged in international aviation markets. Beyond its regulatory functions, the FAA operates and maintains the vast network of air traffic control facilities across the United States, manages the development of the Next Generation Air Transportation System, and conducts research to advance aviation safety and efficiency. The agency's approach combines rigorous technical standards with practical operational considerations, reflecting the diverse and extensive nature of American aviation activities across general aviation, commercial transport, and emerging sectors such as unmanned aerial systems and commercial space transportation.

6.1.2 European Union Aviation Safety Agency (EASA): Europe's Harmonizing Force

The European Union Aviation Safety Agency represents the collective regulatory authority for European Union member states, working to harmonize civil aviation safety standards across its jurisdiction. Established in 2003, EASA has progressively expanded its responsibilities to create a single, high-level regulatory framework encompassing airworthiness, environmental protection, flight crew licensing, and air operations. The agency's foundational Regulation establishes its comprehensive mandate to promote the highest common standards of safety and environmental protection in civil aviation throughout Europe.

EASA's technical standards and certification processes are respected worldwide for their rigorous and systematic approach to safety management. The agency's certification activities cover aircraft type certification, organization approvals, and personnel licensing, with its decisions being valid across all EU member states. EASA also plays a significant role in the international aviation community through its participation in global standard-setting organizations and its bilateral agreements with other aviation authorities. The agency's regulatory philosophy emphasizes risk-based oversight, performance-based regulations, and the integration of new technologies while maintaining the highest safety standards. This approach has positioned EASA as a influential voice in global aviation safety discussions and a key partner in international aviation development initiatives.

6.1.3 Directorate General of Civil Aviation (DGCA): India's Growing Regulatory Presence

The Directorate General of Civil Aviation serves as India's national regulatory body for civil aviation, overseeing all aviation activities within the country's rapidly expanding air transport market. As India has emerged as one of the world's fastest-growing aviation economies, the DGCA's role and responsibilities have expanded correspondingly to meet the challenges of managing this growth while maintaining safety standards.

The agency's functions encompass flight safety oversight, personnel licensing, aircraft airworthiness certification, and operator surveillance. The DGCA develops and enforces civil air regulations tailored to India's specific operational environment and infrastructure challenges while aligning with international standards established by the International Civil Aviation Organization. The agency's certification processes are essential for pilots, engineers, and other aviation professionals seeking to work within India's aviation sector or with Indian-registered aircraft.

In recent years, the DGCA has undertaken significant organizational modernization to enhance its regulatory capabilities, including the implementation of more sophisticated surveillance systems, digital transformation of certification processes, and strengthened safety management systems. The agency's evolving regulatory approach reflects the unique characteristics of the Indian aviation market, including its geographic diversity, operational challenges, and rapid technological adoption, while maintaining alignment with global best practices in aviation safety oversight.

6.1.4 Core Regulatory Functions and Certification Processes

All three regulatory bodies perform similar core functions despite their different jurisdictional contexts, with aircraft certification representing one of their most critical responsibilities. The process for certifying a new aircraft type involves multiple stages of design evaluation, testing, and compliance demonstration to ensure that the aircraft meets all applicable safety standards before entering service. For manufacturers, navigating these certification processes is essential for market access, with many pursuing simultaneous certifications from multiple authorities to enable global operations. Personnel licensing constitutes another fundamental regulatory function, with each authority establishing specific requirements for pilot training, experience, and testing that must be completed before issuing licenses. The licensing frameworks reflect different philosophical approaches to pilot training, ranging from more flexible experience-based models to highly structured training programs with extensive theoretical knowledge requirements. Operator certification represents a third critical function, whereby airlines and other commercial operators must demonstrate their operational competence, maintenance capabilities, and safety management systems before receiving approval to conduct commercial operations. These certification processes typically involve detailed documentation reviews, facility inspections, and operational demonstrations to verify compliance with regulatory standards. Throughout all these functions, the regulatory bodies maintain ongoing surveillance activities to ensure continued compliance after initial certification, conducting regular audits, inspections, and monitoring of safety performance indicators to identify potential issues before they develop into safety hazards.

6.1.5 International Cooperation and Regulatory Harmonization

Despite their distinct regulatory frameworks, the FAA, EASA, and DGCA engage in extensive international cooperation to facilitate global aviation operations and enhance safety standards worldwide. Bilateral Aviation Safety Agreements represent key instruments for this cooperation, establishing frameworks for mutual recognition, technical cooperation, and joint certification activities between regulatory authorities. These agreements help reduce duplication in certification processes, enable more efficient validation of each other's approvals, and promote regulatory alignment in key safety areas.

International Regulatory Cooperation

The International Civil Aviation Organization serves as a global forum for developing international standards and recommended practices that form the basis for much of national regulatory frameworks. While ICAO standards are not directly binding, most states incorporate them into their national regulations, creating a foundation of global consistency in essential safety requirements. The three authorities also participate in various regional and international coordination groups to address emerging safety issues, share safety information, and develop common approaches to new regulatory challenges. This cooperative environment enables authorities to leverage each other's technical expertise, coordinate on complex certification projects involving new technologies, and respond more effectively to safety issues with international implications. Despite this cooperation, differences in regulatory philosophy, implementation timelines, and specific technical requirements continue to present challenges for manufacturers and operators working across multiple jurisdictions, driving ongoing efforts toward greater regulatory harmonization.

6.1.6 Emerging Challenges and Evolving Regulatory Landscapes

The global regulatory bodies face numerous emerging challenges as they adapt to rapid technological changes and evolving industry dynamics. The integration of new aircraft technologies including electric propulsion, advanced composite materials, and increasingly automated systems requires continuous updates to certification standards and methods. These new technologies often challenge traditional regulatory approaches developed for conventional aircraft, necessitating the development of new certification frameworks that can adequately address novel safety considerations while not unduly impeding innovation. The rapid growth of unmanned aircraft systems represents another significant regulatory challenge, requiring the development of entirely new regulatory frameworks to safely integrate these systems into existing airspace structures.

Environmental concerns including aircraft emissions and noise pollution are increasingly influencing regulatory priorities, with authorities developing new standards and promoting technologies that reduce aviation's environmental footprint. The globalization of aviation manufacturing and maintenance has complicated oversight responsibilities, requiring enhanced international cooperation and new approaches to managing safety across extended supply chains. Cybersecurity has emerged as a critical safety consideration as aircraft and air traffic management systems become more digitally connected and potentially vulnerable to cyber threats. All these challenges are occurring against a backdrop of continued air traffic growth in many regions, increasing the stakes for effective safety oversight and creating pressure on regulatory resources. In response, the authorities are increasingly adopting risk-based oversight approaches, leveraging data analytics to identify emerging risks, and exploring new regulatory models that can more effectively address the pace of technological change while maintaining fundamental safety standards.

- **Managing Technological Disruption**: Regulatory bodies must continuously adapt their frameworks to address emerging technologies including electric propulsion, advanced automation, and new materials that challenge traditional certification approaches.
- **Integrating Unmanned Systems**: The rapid proliferation of drones requires developing comprehensive regulatory frameworks for their safe integration into shared airspace alongside conventional aircraft.

- **Addressing Environmental Pressures:** Increasing focus on sustainability demands new regulations governing aircraft emissions, noise pollution, and promoting environmentally friendly technologies.
- **Ensuring Cybersecurity:** As aircraft systems become more connected, regulators must establish robust cybersecurity standards to protect against potential threats to aircraft systems and operations.
- **Managing Global Supply Chains:** The increasingly international nature of aircraft manufacturing requires sophisticated oversight of extended supply chains and production facilities across multiple jurisdictions.
- **Harmonizing International Standards:** Despite cooperation, differences in regulatory requirements across jurisdictions continue to challenge manufacturers and operators seeking global market access.
- **Maintaining Safety During Rapid Growth:** Particularly in expanding markets, regulators must ensure safety standards keep pace with rapid increases in air traffic and industry expansion.
- **Modernizing Regulatory Frameworks:** Traditional regulatory approaches must evolve to address new operational concepts including urban air mobility and increasingly automated aircraft operations.
- **Workforce Development:** Ensuring adequate technical expertise within regulatory organizations to effectively oversee rapidly advancing technologies represents an ongoing challenge.
- **Balancing Safety and Innovation:** Regulators must maintain stringent safety standards while avoiding unnecessary impediments to technological innovation and industry development.

6.1.7 Impact on Industry Stakeholders and Global Aviation

The regulations and certification processes established by the FAA, EASA, and DGCA profoundly influence all participants in the aviation industry, from aircraft manufacturers to airlines and service providers. For manufacturers, regulatory requirements directly shape design decisions, development timelines, and certification strategies, with significant implications for development costs and market access. The choice of which regulatory framework to pursue for initial certification can influence global market potential, particularly for new aircraft programs targeting international operations. Airlines and operators must navigate complex regulatory requirements governing their operations, maintenance practices, and personnel training, with compliance representing a significant operational consideration and cost factor. The differences between regulatory frameworks can affect fleet planning decisions, operational procedures, and crew training requirements, particularly for operators conducting international operations across multiple regulatory jurisdictions. Maintenance organizations must obtain appropriate certifications from the relevant authorities and maintain strict compliance with continuing airworthiness requirements.

For aviation professionals, the specific licensing requirements of each authority influence career planning, training choices, and geographic mobility, with many professionals maintaining multiple certifications to enhance their employment opportunities. The regulatory landscape also significantly influences the development and adoption of new technologies, as manufacturers must consider certification pathways and timelines when making technology investment decisions. Overall, the regulatory frameworks established by these bodies create the fundamental safety infrastructure that enables modern aviation while simultaneously shaping industry structure, operational practices, and technological development through their specific requirements and approval processes.

6.2 CERTIFICATION PATHWAYS FOR ELECTRIC AIRCRAFT

Electric aircraft certification is not a wholesale reinvention of airworthiness; it is the methodical application of established safety processes to new energy and propulsion technologies. Authorities have created workable pathways—SCVTOL and poweredlift for eVTOL, Part/CS23 for many fixedwing electrics—supplemented by special conditions and modern system standards.

Success hinges on integrating batteries, highvoltage systems, and distributed propulsion into a safety case that stands up under test, with cybersecurity and continued airworthiness built in from the start. Programs that engage regulators early, ground novel claims in validated models and data, and stage operational ambitions will cross the certification finish line sooner—and with aircraft that are safer, quieter, and more reliable for the communities they aim to serve.

6.2.1 Regulatory Landscape

In the United States, the FAA certifies aircraft, engines, and propellers under 14 CFR Part 21, using specific airworthiness standards such as Part 23 (normal category airplanes), Parts 27/29 (rotorcraft), and, for eVTOL, "special class" poweredlift criteria under §21.17(b). The FAA publishes applicantspecific airworthiness criteria in the Federal Register for new poweredlift types, while operations typically fall under Part 91 for private use or Part 135 for air taxi/charter. EASA plays a similar role in Europe, with CS23/27/29 and a dedicated Special Condition—VTOL (SCVTOL) for multirotor and liftpluscruise vehicles. Other authorities—UK CAA, Transport Canada, ANAC Brazil, CASA Australia—align broadly with FAA/EASA through bilateral agreements, though details and timelines vary.

For propulsion, regulators are converging on integrated certification of electric propulsion systems within the aircraft type rather than as standalone "engines" under legacy rules. EASA's special conditions for electric/hybrid propulsion and FAA issue papers establish requirements for functional safety, thermal management, electrical protection, and continued airworthiness. Batteries, highvoltage buses, and power electronics trigger additional standards, test methods, and design assurance expectations.

6.2.2 Building Blocks: From Type to Operations

Certification unfolds in layers. Type certification (TC) establishes that a design meets airworthiness standards; it begins with agreement on the certification basis (e.g., Part 23 at a given amendment, SCVTOL) and means of compliance, captured in issue papers (FAA G1, G2) or EASA certification review items. After TC, a production certificate (PC) or production organization approval (POA) authorizes serial manufacture under a quality system that ensures conformity. Individual aircraft receive airworthiness certificates (standard or special) based on conformity inspections.

Operational approvals then govern how the aircraft is used. For passenger services, an Air Operator Certificate (AOC) and operations specifications (OpSpecs) under Part 135 (FAA) or equivalent EASA approvals define crew training, maintenance programs, performance limitations, and routes. For eVTOL, the FAA has introduced poweredlift pilot and maintenance qualifications; EASA expects conversion training specific to VTOL handling qualities and energy management. The result is a dual track: airworthiness of the machine and competency of the organization that flies and maintains it.

6.2.3 System Safety and Development Standards

Electric aircraft integrate novel systems that must satisfy classical safety objectives. The development framework remains ARP4754A for system engineering and ARP4761A for safety assessment, tracing from functional hazard assessments (FHA) to system safety assessments (SSA) with quantitative failure probability targets tied to severity. Software is qualified to DO178C at appropriate Design Assurance Levels (DAL), with tool and modelbased supplements (DO330/331) as needed. Complex electronic hardware follows DO254. Environmental qualification uses DO160G/H for temperature, vibration, EMI/EMC, lightning, and power quality, often with more stringent assumptions for highvoltage transient behavior.

Cybersecurity now figures prominently. DO326A/ED202A (airworthiness security process), DO355 (information security guidance), and DO356A (airworthiness security methods) guide threat assessment, protection, and verification so that safetycritical functions cannot be compromised. Integrated modular avionics and partitioning (DO297) help isolate functions with different assurance levels and reduce crosscoupling between noncritical and critical software.

6.2.4 Batteries and HighVoltage Energy Storage

Rechargeable lithium batteries are certified under DO311A for safety, abuse testing, and thermal runaway behavior. Aviationgrade packs must demonstrate cell nonpropagation or controlled propagation, venting paths that avoid occupant exposure, robust containment, and maintainability. Protections include independent sensing and contactors, faulttolerant battery management systems (BMS), and proven algorithms for state of charge/energy estimation. Electrical wiring interconnection systems (EWIS) practices help minimize arcing and manage creepage/clearance at hundreds of volts.

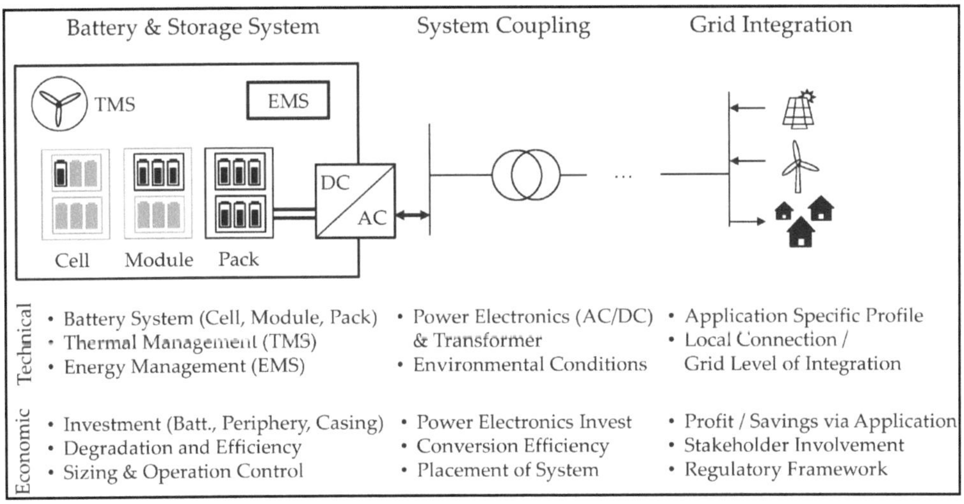

Lithium-Ion Battery Storage for the Grid

Thermal management is part of the certification story: liquid jackets, heat exchangers, and preconditioning must keep cells within safe envelopes across worstcase missions and ambient extremes. Charge interfaces and ground equipment add a boundary condition; while not part of the aircraft TC per se, their interplay with battery safety (e.g., permissible charge rates versus SOH and temperature) is scrutinized through installation and operational approvals.

6.2.5 Electric Propulsion and Power Distribution

Electric motors, inverters, and propulsors are approved as part of the aircraft's propulsion system. Requirements address performance, structural integrity, and functional safety during normal and failure conditions—loss of phase, loss of cooling, singleevent effects, controller failures—and their impact on thrust and controllability. Distributed propulsion introduces redundancy but also new failure combinations; certification demands demonstration that the aircraft can handle motor/inverter failures without loss of control, including propulsorout climb or hover margins. Highvoltage DC buses require protection against arc faults, ground faults, and over/undervoltage conditions. Segmented buses with contactors, current limiting, and fast fault detection isolate faults without collapsing power to critical loads. Lightning indirect effects and highintensity radiated fields (HIRF) analyses ensure insulation and bonding prevent hazardous transients. In multirotor eVTOL, synchronization and control authority across many propulsors are central to handling qualities and failure response.

6.2.6 eVTOL and PoweredLift Specifics

eVTOL certification layers in VTOLspecific risks. Under EASA SCVTOL or FAA poweredlift criteria, applicants must show safe transition between hover and wingborne flight, adequate control margins with elevated downwash and gusts, and performance in confined area operations. Hover/transition reliability, handling qualities, and energy reserves (often expressed as minutes of hover or alternate profiles) are core.

Loss of thrust scenarios—single or multiple propulsors—must be survivable via redundancy and control, or mitigated by ballistic recovery systems where appropriate and approved. Noise certification and community acceptance are intertwined. While conventional Part 36 metrics apply to many aircraft, authorities are shaping eventbased and urbancontext metrics for VTOL operations. Flight testing must characterize tonal content, operational profiles, and procedures that minimize noise footprints, feeding both certification and operational approvals.

6.2.7 Flight Test, Ground Test, and Means of Compliance

Means of compliance mix analysis, simulation, bench testing, and flight demonstrations. Early in the program, component and ironbird rigs validate powertrain controls, fault responses, and thermal behavior. Structural tests validate crashworthiness, bird strike resistance, and rotor/propeller integrity. Lightning and EMI tests validate shielding and bonding.

Flight tests expand the envelope methodically, proving performance, control margins with degraded systems, energy prediction accuracy, and conformance to noise and reserve requirements. For batteries, abuse tests and propagation trials are conducted at pack level with representative installation features. Regulators expect traceable models—validated by test—to support certification credit. Digital twins used for energy management and failure simulation gain credibility when corroborated by telemetry across varied conditions. Clear test hazard analyses and safety cases are mandatory, particularly for publicarea eVTOL trials.

6.2.8 Continued Airworthiness and Maintenance

Certification does not end at the type certificate. Instructions for Continued Airworthiness (ICA) define inspection intervals, replacement thresholds, and maintenance procedures. For electric aircraft, ICAs must cover battery health metrics (state of health, impedance rise), endoflife criteria, module replacement processes, and highvoltage safety protocols.

Motor/inverter cooling circuits, dielectric inspections, and connector integrity become routine checks. Operators' reliability programs and MSG3 analyses inform escalation or deescalation of tasks as fleet data accumulates. Software updates and configuration control demand rigorous change management. Overtheair updates may be allowed under strict controls; each release must preserve safety and certification baselines (including cybersecurity). Event reporting and service bulletins feed continuous improvement, and authorities may require inservice monitoring for early electric types to confirm assumptions made during certification.

6.2.9 Unmanned and Optionally Piloted Electric Aircraft

Unmanned electric aircraft follow different pathways depending on size and risk. Small drones operate under rules like FAA Part 107 or EASA's Open/Specific categories; higherrisk BVLOS and cargo operations use Specific (SORAbased) or Certified categories with airworthiness approvals approaching crewed rigor. Large cargo drones entering controlled airspace face requirements similar to crewed commuter aircraft, including design assurance, detectandavoid, C2 link reliability, and maintenance programs. Optionally piloted conversions must satisfy both crewed and unmanned safety objectives, often staged through phased operational approvals.

6.2.10 Global Harmonization and Timelines

Bilateral agreements and working groups aim to reduce duplicative effort, but applicants should anticipate jurisdictionspecific interpretations. Early alignment on certification basis and means of compliance prevents late rework. Timelines are influenced by novelty—the first of type in a category will set precedents and often move slower. Many programs derisk via incremental certifications: start with piloted, dayVMC operations; expand to IFR, icing approval, and more demanding profiles; then pursue automation or higher utilization based on inservice evidence.

6.2.11 Practical Moves to DeRisk Electric Aircraft Certification

- Engage regulators early to agree on certification basis, special conditions, and means of compliance; lock G1/G2 scope before design freezes.

- Build a safety case around ARP4754A/ARP4761A with clear FHA/PSSA/SSA artifacts; tie requirements to quantitative energy and propulsion hazards.
- Treat batteries as certificated systems: DO311A test plans for abuse and nonpropagation, robust BMS architecture, and verifiable SOE accuracy.
- Design highvoltage architecture for fault tolerance: fast arc/ground fault detection, segmented buses, and safe reconfiguration under failures.
- Validate propulsion failure response on ironbirds and flight test: propulsorout handling, degraded hover/transition margins, and thermal derates.
- Develop cybersecurity from day one (DO326A/355/356A); secure boot, signed updates, and segmented networks across avionics and charging links.
- Prove noise and community profiles early; use data to refine approach/transition procedures and support operational approvals and public acceptance.
- Stand up a production quality system early (POA/PC path); ensure configuration control for batteries, software, and serialized powertrain parts.
- Write ICAs that are practical: battery SOH criteria, safe HV handling, cooling system maintenance, and datadriven reliability programs.
- Plan an incremental ops roadmap: dayVMC to IFR, expand envelopes with service experience, and align pilot/maintainer training with each step.

6.3 SAFETY CHALLENGES WITH BATTERIES AND HIGH-VOLTAGE SYSTEMS

6.3.1 Fundamental Battery Safety Hazards

Battery systems present unique safety challenges stemming from their fundamental operating principles and the concentration of chemical energy in confined spaces. The electrochemical reactions that enable energy storage and release can become uncontrolled under certain conditions, leading to rapid heat generation, gas evolution, and potential explosive failure.

Modern high-energy batteries operate near their stability limits to maximize performance, leaving minimal safety margins. Internal defects, whether from manufacturing variations, mechanical damage, or degradation over time, can create conditions where normal protective mechanisms fail to prevent hazardous events. The complexity of battery pack construction multiplies individual cell risks through potential cascade failures. A single cell experiencing thermal runaway can propagate failure throughout an entire pack, as heat and ejected materials compromise neighboring cells.

This propagation mechanism transforms localized failures into system-wide catastrophes, particularly challenging in large format applications like electric vehicles or stationary storage systems. The sealed nature of battery packs complicates early detection of developing problems, often allowing conditions to deteriorate significantly before external symptoms manifest. Additionally, the DC nature of battery systems creates unique electrical hazards including sustained arc faults that conventional AC protection devices cannot adequately address.

6.3.2 High-Voltage System Risks

High-voltage systems introduce distinct hazards beyond those inherent to batteries alone, with potentially lethal consequences for inadequately protected personnel. Voltage levels in modern electric vehicles and energy storage systems commonly exceed 400 volts, with some applications reaching 800 volts or higher to improve efficiency. These voltages far exceed human safety thresholds, where even brief contact can cause cardiac arrest, severe burns, or fatal electrocution. The silent, invisible nature of electrical hazards makes them particularly insidious, as workers may not perceive danger until contact occurs.

Arc flash represents one of the most severe high-voltage hazards, capable of generating temperatures exceeding 35,000 degrees Fahrenheit and blast pressures that can throw personnel across rooms. The DC characteristics of battery systems create particularly persistent arcs that don't self-extinguish at current zeros like AC arcs, requiring sophisticated interruption techniques. Stored energy in capacitors and inductors within high-voltage systems can maintain dangerous potentials even after main disconnects are opened, creating risks during maintenance and emergency response. The proliferation of high-voltage systems into consumer applications like personal vehicles introduces these industrial-grade hazards into everyday environments where users lack specialized safety training.

6.3.3 Thermal Runaway and Fire Hazards

Thermal runaway remains the most dramatic and challenging safety concern in battery systems, particularly for lithium-ion chemistries. This phenomenon occurs when heat generation within a cell exceeds heat dissipation capacity, creating a positive feedback loop that accelerates chemical decomposition reactions. Initial triggers include overcharging, internal short circuits, manufacturing defects, or external heating, but once initiated, thermal runaway becomes self-sustaining through exothermic decomposition of electrode materials and electrolytes. Temperatures can rapidly exceed 1000°C, with violent ejection of flammable gases and molten materials.

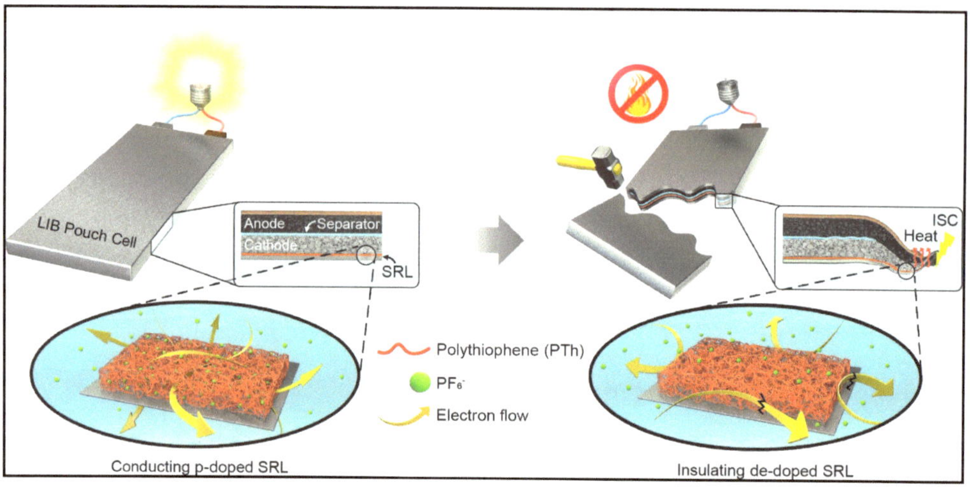

Heat-Induced Runaway and Fire Dangers

Battery fires present unique suppression challenges those conventional firefighting methods often cannot address effectively. Water application may cool external surfaces but cannot penetrate sealed battery enclosures to address the source of heat generation. Additionally, water contact with certain battery chemistries can generate hydrogen gas or toxic compounds. The stranded energy within damaged battery systems can reignite fires hours or days after apparent suppression, requiring extended monitoring and specialized handling procedures. Large-scale battery installations multiply these challenges, as the heat and toxic gas generation from massive battery fires can overwhelm traditional fire protection systems and create area-denial conditions preventing effective emergency response.

6.3.4 Chemical and Environmental Hazards

Beyond thermal and electrical risks, battery systems contain numerous chemical hazards that pose acute and chronic health risks. Electrolytes typically consist of organic solvents and lithium salts that are flammable, corrosive, and toxic. During normal operation, these materials remain sealed within cells, but mechanical damage, thermal events, or age-related degradation can release them into the environment. Fluorinated compounds common in electrolytes can decompose into hydrogen fluoride, an extremely dangerous gas that causes severe respiratory damage and deep tissue burns even at low concentrations.

Environmental considerations extend throughout the battery lifecycle, from raw material extraction through end-of-life disposal. Mining operations for lithium, cobalt, nickel, and other battery materials often involve environmentally destructive processes and hazardous working conditions. Manufacturing processes utilize toxic solvents and generate hazardous waste streams requiring careful management. End-of-life batteries contain valuable materials worth recovering but also hazardous compounds requiring specialized recycling processes. Improper disposal can lead to soil and groundwater contamination, while informal recycling in developing nations often exposes workers to severe health hazards without adequate protection.

6.3.5 Manufacturing and Production Safety

Battery manufacturing environments concentrate multiple hazards including flammable solvents, reactive materials, and high-energy processes that demand rigorous safety protocols. The production of electrode materials involves handling fine powders that pose explosion risks when suspended in air, requiring specialized ventilation and inerting systems. Electrolyte filling operations must manage flammable, moisture-sensitive materials in controlled atmospheres to prevent reactions with ambient humidity. Formation cycling, where cells receive their first charge, generates gases that must be safely vented while monitoring for anomalous behavior indicating defective cells.

Quality control throughout manufacturing critically impacts downstream safety, as microscopic defects can create failure points months or years later. Contamination with metallic particles, separator defects, or electrode coating variations can create internal short circuit paths. However, detecting these defects challenges current inspection technologies, as many failure modes only manifest under specific operating conditions. The pressure for cost reduction and production scaling sometimes conflicts with thorough quality assurance, particularly as battery manufacturing expands into regions with developing industrial safety cultures. Automation reduces human exposure to manufacturing hazards but introduces new risks related to high-speed machinery and the concentration of hazardous materials.

6.3.6 Transportation and Storage Challenges

The classification of lithium batteries as dangerous goods creates complex logistics challenges for safe transportation across all modes. Regulations limit quantities, require specialized packaging, and mandate specific handling procedures to prevent incidents during shipping. The potential for cargo fires on aircraft led to restrictions on battery shipments and requirements for fire-resistant containers.

Maritime transport faces similar concerns, with several ship fires attributed to battery cargo highlighting the catastrophic potential of large-scale incidents in confined vessels far from emergency response resources. Storage facilities must address both routine safety requirements and emergency scenarios involving large quantities of batteries. Thermal management becomes critical, as elevated temperatures accelerate degradation and increase failure risks. State-of-charge management during storage affects both safety and battery longevity, with partially charged states generally offering the best compromise. Segregation strategies must consider preventing cascade failures while maintaining practical warehouse operations. Automated monitoring systems increasingly deploy sensors for temperature, smoke, and off-gas detection, but the rapid progression from detection to serious incident challenges response times.

6.3.7 Emergency Response Considerations

First responders face unprecedented challenges when dealing with battery and high-voltage system incidents. Traditional firefighting training inadequately prepares personnel for the unique characteristics of battery fires, electrical hazards from damaged high-voltage systems, and toxic gas exposures. Electric vehicle accidents introduce scenarios where victims may be trapped in proximity to damaged high-voltage components, requiring specialized extrication procedures and protective equipment. The delayed nature of some battery failures means scenes that appear safe can become hazardous hours after initial incidents.

Development of specialized response protocols continues evolving as experience accumulates from real-world incidents. These protocols must address immediate life safety, scene stabilization, and long-term hazard mitigation while protecting emergency personnel. Equipment needs include thermal imaging cameras for identifying hot spots, gas monitors for toxic compounds, and specialized personal protective equipment rated for electrical and chemical hazards. Training programs must reach broadly across emergency services, as even rural departments may encounter electric vehicles or energy storage systems. Coordination between manufacturers, emergency responders, and regulatory agencies remains essential for developing effective response strategies.

6.3.8 Design Solutions and Safety Systems

Engineering solutions for battery and high-voltage safety employ multiple layers of protection addressing different failure modes and consequences. Cell-level safety devices include pressure relief vents, positive temperature coefficient resistors, and shutdown separators that interrupt current flow when triggered by abnormal conditions.

Pack-level protections add fusing, contactors, and isolation monitoring to prevent and mitigate failures. Battery management systems continuously monitor voltage, current, and temperature parameters while implementing protective actions when limits are exceeded. Advanced design approaches increasingly emphasize inherent safety through material selection and system architecture rather than relying solely on active protection systems. Solid-state batteries promise improved safety by eliminating flammable liquid electrolytes, though manufacturing challenges currently limit commercialization. Physical segregation strategies using thermal barriers and propagation-resistant materials limit cascade failures. Cooling system designs must balance normal thermal management needs with emergency heat rejection capabilities during abnormal events. Integration of multiple sensor types and artificial intelligence-based anomaly detection enables earlier intervention before conditions deteriorate to dangerous levels.

6.4 CYBERSECURITY AND DIGITAL FLIGHT SYSTEMS

Cybersecurity and digital flight systems are now inseparable elements of modern aviation. As aircraft evolve into flying data centers, safeguarding their electronic systems becomes as vital as maintaining their mechanical integrity. The integration of cybersecurity into every phase of flight operation—from software design to real-time monitoring—ensures that the aviation industry can withstand emerging digital threats without compromising safety or efficiency. With advances in artificial intelligence, quantum encryption, and international cooperation, the future of aviation will be defined not only by how far and fast we can fly, but also by how securely we can navigate the skies.

6.4.1 Evolution of Digital Flight Systems

Digital flight systems, often referred to as avionics, have evolved from analog instrumentation to sophisticated electronic and software-driven architectures. Modern aircraft employ systems such as Fly-by-Wire (FBW), Flight Management Systems (FMS), Automatic Dependent Surveillance-Broadcast (ADS-B), and Electronic Flight Bags (EFBs) to optimize navigation, control, and performance. These systems rely on high-speed data processing and extensive network connectivity to enhance flight precision, reduce pilot workload, and improve situational awareness. However, as these systems become more interconnected, they also become potential entry points for cyber intrusions, necessitating multilayered defense mechanisms and constant system monitoring.

6.4.2 Cyber Threat Landscape in Aviation

The cybersecurity threat landscape in aviation is both dynamic and complex. Potential attackers include state-sponsored actors, criminal organizations, and hacktivists seeking to exploit system vulnerabilities. Threats range from data breaches and ransomware attacks to manipulation of navigation signals and denial-of-service (DoS) disruptions. Digital flight systems, because of their reliance on real-time communication between air and ground networks, are particularly susceptible to interference or data spoofing. For example, attacks on ADS-B signals could mislead air traffic controllers by providing false positional data, while compromised FMS software might affect flight paths or fuel calculations. Therefore, protecting these systems requires both proactive threat intelligence and resilient architecture design.

Digital Threat Environment in Aviation

6.4.3 Integration of Cybersecurity in Flight Operations

To safeguard modern aviation, cybersecurity must be integrated throughout the lifecycle of flight systems—from design and manufacturing to operations and maintenance. Aircraft manufacturers now implement security-by-design principles, embedding encryption protocols and intrusion detection systems into avionics from the earliest development stages. Airlines employ Security Operations Centers (SOCs) to monitor digital networks continuously and respond to incidents in real time. Moreover, maintenance teams use digitally signed software updates to prevent unauthorized code injections. The International Civil Aviation Organization (ICAO) and other agencies have also introduced frameworks to standardize cybersecurity practices, ensuring that safety remains paramount across all stakeholders.

6.4.4 Components of Digital Flight Systems and their Vulnerabilities

Digital flight systems encompass a wide range of subsystems, each with its unique cybersecurity challenges. The Flight Management System relies on GPS and navigation databases, making it susceptible to data spoofing. The Fly-by-Wire control system, which replaces mechanical linkages with electronic signals, demands secure communication channels between cockpit controls and actuators. Aircraft Communication Addressing and Reporting System (ACARS) messages, used for transmitting operational information, are vulnerable if left unencrypted.

Additionally, Wi-Fi networks onboard commercial flights, though separated from critical avionics, can become potential gateways if improperly secured. Hence, layered defense, segmentation, and real-time monitoring are crucial for mitigating risks across these interconnected systems.

Table: Traditional vs. Digital Flight Systems

Feature	Traditional Flight Systems	Digital Flight Systems
Control Mechanism	Mechanical and analog controls	Electronic Fly-by-Wire systems
Communication	Limited radio-based exchange	High-speed data and satellite links
Maintenance	Manual inspection and logging	Automated diagnostics and predictive analytics
Vulnerability	Physical tampering	Cyber intrusion and data manipulation
Security Approach	Basic mechanical safeguards	Advanced encryption and network security

6.4.5 Regulatory Frameworks and International Standards

Cybersecurity in aviation operates under a growing web of international regulations and guidelines. The ICAO Aviation Cybersecurity Strategy provides a global framework that promotes collaboration and risk management across nations. The European Union Aviation Safety Agency (EASA) and the Federal Aviation Administration (FAA) have introduced standards requiring cybersecurity risk assessments during aircraft certification. Additionally, the RTCA DO-326A/ED-202A standard outlines processes for developing secure airborne systems. Compliance with such frameworks ensures that manufacturers, airlines, and service providers align their cybersecurity measures with global safety expectations. These regulations are continuously updated to address emerging threats, ensuring that aviation remains a step ahead of cyber adversaries.

6.4.6 Cybersecurity Technologies in Aviation Defense

The protection of digital flight systems relies on an integrated suite of cybersecurity technologies. Encryption protocols safeguard data in transit between aircraft and ground stations. Intrusion Detection Systems (IDS) and Intrusion Prevention Systems (IPS) continuously monitor network traffic to identify suspicious behavior. Blockchain-based authentication is being explored to ensure the integrity of maintenance records and software updates. Meanwhile, Artificial Intelligence (AI) and Machine Learning (ML) enhance anomaly detection by identifying patterns that deviate from normal system behavior. Redundancy and isolation—key principles in aviation—are being adapted to cybersecurity, ensuring that a breach in one system does not cascade across the entire aircraft network.

6.4.7 Human Factors and Training in Cybersecurity

While technology forms the backbone of cybersecurity, human factors remain equally crucial. Pilots, engineers, and air traffic personnel must be trained to recognize and respond to cyber threats. Many incidents result from social engineering, where attackers exploit human error rather than technical vulnerabilities. Therefore, airlines and aviation authorities are investing in cyber awareness training programs and simulation exercises to enhance preparedness. Maintenance personnel are also taught secure software update protocols and password management practices. Building a culture of cyber vigilance across all aviation employees strengthens the industry's resilience against evolving threats and minimizes the impact of potential breaches.

6.4.8 Aviation Cybersecurity

- Increasing system complexity and interconnectivity
- Lack of uniform cybersecurity standards across jurisdictions
- Difficulty in patching and updating airborne systems
- Limited bandwidth for secure communication in flight
- Integration of legacy systems with modern avionics
- Insider threats within airlines or maintenance facilities
- Supply chain vulnerabilities in software components
- Dependence on third-party data providers and satellite links
- Balancing security with operational efficiency
- Rapid evolution of cyber-attack techniques and malware

6.4.9 Emerging Concepts: Artificial Intelligence and Quantum Security

Artificial Intelligence (AI) is playing a transformative role in aviation cybersecurity. AI-driven monitoring systems can analyze vast amounts of telemetry data in real time, detecting anomalies that might indicate an attack. Predictive algorithms forecast potential vulnerabilities based on historical data and system usage patterns. Moreover, quantum encryption technologies are being researched to provide virtually unbreakable communication channels for aircraft systems. Quantum key distribution (QKD) could revolutionize secure data exchange between aircraft and satellites, providing next-generation protection against cyber espionage. As these technologies mature, they will form the foundation of a more secure digital aviation ecosystem.

6.4.10 Case Studies of Cyber Incidents in Aviation

A few incidents have highlighted the urgency of cybersecurity in aviation. In 2015, researchers demonstrated that vulnerabilities in in-flight entertainment systems could theoretically allow limited access to non-critical networks.

Similarly, ransomware attacks on airline ground systems have disrupted flight schedules and passenger data systems. While no major cyberattack has yet compromised flight-critical systems, these incidents underscore the need for preemptive action. The aviation industry has responded with stricter segmentation between passenger and avionics networks, enhanced monitoring, and the adoption of cybersecurity best practices across operational domains.

6.4.11 Collaboration and Information Sharing

Given the global nature of aviation, cybersecurity cannot be managed in isolation. Effective defense requires international collaboration among airlines, manufacturers, regulators, and cybersecurity agencies. Information sharing platforms such as the Aviation Information Sharing and Analysis Center (A-ISAC) enable stakeholders to exchange real-time threat intelligence and mitigation strategies. Collaborative exercises and cybersecurity summits allow for cross-sectoral learning and coordinated responses to incidents. Governments and private industry are increasingly partnering to create public-private cybersecurity frameworks, ensuring that national security and commercial safety objectives are aligned in the face of evolving cyber challenges.

6.5 AIR TRAFFIC MANAGEMENT FOR ELECTRIC AND UAM OPERATIONS

6.5.1 Operational Landscape and Core Challenges of Urban Airspace

Urban Air Mobility is envisioned as a transport system for moving people and goods within urban areas using electric vertical takeoff and landing aircraft. The operational concept for this new class of operations continues to evolve, with regulatory bodies developing comprehensive operational concepts to integrate these new vehicles into existing airspace and vertiport infrastructure. The core challenge lies in bridging the gap between theoretical frameworks and their practical, safe application in complex urban environments.

Key operational questions that must be answered include how to manage conflict avoidance between piloted, remotely piloted, and automated vehicles; how to ensure safe operations near airports and in areas with limited satellite reception; and how to maintain target levels of safety amidst non-cooperative airspace users like birds or unauthorized drones. The density of operations envisioned for urban air mobility creates a traffic management problem that is fundamentally different in scale and kind from traditional aviation, requiring new thinking about airspace structure, separation standards, and contingency management.

- **Managing High-Density, Low-Altitude Operations**: Urban air traffic management systems must safely manage a volume of aircraft orders of magnitude greater than current systems handle, all within constrained urban corridors and at lower altitudes where traditional air traffic control has limited experience.
- **Integration with Traditional Air Traffic Control**: A primary challenge is seamlessly integrating urban air mobility operations, particularly in terminal airspace near airports, without disrupting the flow of existing commercial aviation and while maintaining current safety standards.
- **Ensuring Safety and Security**: Guaranteeing the safety of passengers and people on the ground is paramount, requiring comprehensive approaches to system safety, cybersecurity threats, and physical risks from system failures or adverse weather conditions.
- **Real-Time Data Exchange and Communication**: Effective management requires a high-speed, low-latency network for the seamless transmission of mission-critical information, including flight intentions, trajectory modifications, and contingency procedures.
- **Scalable Conflict Detection and Resolution**: Traditional methods like notices to airmen are insufficient for automated, high-tempo operations, necessitating automated, real-time deconfliction systems that can handle hundreds of simultaneous operations.
- **Vertiport Capacity and Ground Operations**: The physical infrastructure of vertiports, including takeoff, landing, and charging pads, can become a primary bottleneck, limiting overall network throughput and requiring sophisticated ground management systems.
- **Public Acceptance and Noise Mitigation**: Community concerns about noise, privacy, and safety must be addressed through quiet vehicle design, careful route planning, and transparent communication about operational patterns and safety records.
- **Economic Viability and Regulatory Uncertainty**: High investment costs and an evolving regulatory landscape create economic uncertainty, which can hinder the widespread implementation of urban air mobility services and infrastructure.
- **Beyond Visual Line of Sight Operations**: Scaling urban air mobility requires routine beyond visual line of sight flights, which depend on robust detect-and-avoid technology and regulatory approval for complex operations in populated areas.

- **Environmental Integration**: Urban air mobility operations must be conducted in a way that minimizes their environmental footprint, aligning with broader sustainability goals for cities and ensuring compatibility with urban ecosystems.

6.5.2 Evolving Regulatory Frameworks and Key Initiatives

The regulatory landscape for urban air mobility traffic management is rapidly evolving, with leading aviation authorities developing new frameworks to enable safe operations. In the United States, the Federal Aviation Administration is pioneering the Unmanned Aircraft System Traffic Management ecosystem, conceived as a collaborative, distributed network where operators and service providers use highly automated systems to manage operations rather than relying on direct voice communication with air traffic controllers.

The administration's approach includes establishing approval processes and conducting ongoing operational evaluations that allow the agency to assess industry-proposed capabilities and grant approvals for complex operations like package delivery and passenger transport. Simultaneously, in Europe, the European Union Aviation Safety Agency has established the U-space regulatory framework, recognized by industry as a solid foundation toward the vision for a fully digital European sky. These frameworks are not static; they are being stress-tested and refined through real-world pilot projects and sophisticated simulations. Various projects in Europe have defined U-space corridors in major cities to safely link vertiports while avoiding airport approach paths and restricted airspace. These initiatives highlight a global effort to build a regulatory foundation that is performance-based, technology-agnostic, and fosters sustainable competition while maintaining the highest safety standards.

6.5.3 Technological Pillars of Next-Generation Air Traffic Management

The operational and regulatory ambitions for urban air mobility depend entirely on a suite of advanced technologies that form the backbone of modern unmanned traffic management systems.

Distributed service architecture enables multiple service suppliers to interoperate in a collaborative ecosystem, creating a federated network of service suppliers using discovery and synchronization services to coordinate airspace access. Detect and avoid systems provide the critical capability for aircraft to sense and avoid collisions with other airspace users, incorporating onboard perception systems using various transponders, radar, and cameras complemented by ground-based surveillance networks. Artificial intelligence and machine learning technologies power predictive trajectory modeling and automated conflict resolution, using advanced probabilistic aircraft trajectory prediction to assess urban air mobility route feasibility and manage complex airspace interactions. Automated deconfliction systems resolve potential conflicts between flight paths in real-time through strategic deconfliction services and dynamic rerouting capabilities.

Communication and data links facilitate high-speed, low-latency data exchange for command and control, typically using internet-based application programming interfaces connecting service providers to central flight information management systems. These technological components work together to create a robust, scalable system capable of managing the complexity of urban air operations while maintaining safety standards equivalent to traditional aviation.

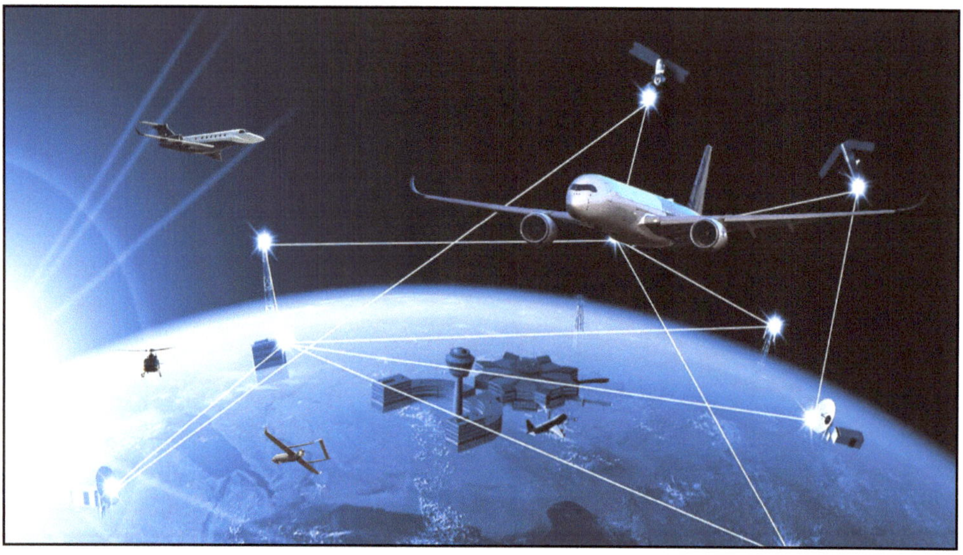

Cutting-Edge Air Traffic Management Systems

6.5.4 Airspace Structure and Operational Concepts

The design of airspace structure for urban air mobility operations represents a critical element in the overall traffic management system. Concepts under development include the establishment of urban air mobility corridors that provide structured routes through dense urban environments, similar to highways in the sky. These corridors must be designed to accommodate different performance characteristics of various aircraft types while maintaining safe separation from buildings, terrain, and other airspace users.

The operational concepts also include various layers of airspace usage, with different altitude bands allocated for specific types of operations such as transit routes, vertiport approaches, and emergency procedures. Dynamic airspace reconfiguration capabilities will be essential to respond to changing weather conditions, traffic demand patterns, and unexpected contingencies. The integration of vertiports into the urban fabric requires careful consideration of approach and departure paths, noise abatement procedures, and ground handling operations. These airspace designs must also incorporate interfaces with traditional air traffic control sectors to ensure seamless transitions between urban air mobility operations and conventional aviation, particularly in terminal areas around major airports. The development of these operational concepts involves extensive simulation and modeling to validate safety, efficiency, and capacity before implementation in real-world environments.

6.5.5 Implementation Challenges and Integration Pathways

The implementation of comprehensive air traffic management for urban air mobility faces numerous challenges that must be addressed through phased integration pathways. The initial phases focus on limited operations in less complex environments, gradually expanding to more dense urban operations as technology and procedures mature. A key challenge involves developing certification standards for the aircraft themselves, along with their avionics and communication systems, to ensure they can operate safely within the proposed management frameworks. The integration of these new systems with existing air traffic control infrastructure requires careful planning to avoid disruption to current aviation operations while creating capacity for new urban air mobility services.

Another significant challenge involves developing contingency procedures for various failure scenarios, including communication loss, navigation degradation, and emergency situations requiring priority handling. The human factors aspects of these systems, whether involving remote pilots, air traffic managers, or automated systems operators, require thorough understanding and integration into the overall system design. Additionally, the business models for service providers, infrastructure operators, and air traffic management services need to be developed to ensure economic sustainability while maintaining safety as the highest priority. These implementation challenges require close collaboration between regulatory authorities, industry stakeholders, and research organizations to develop solutions that enable the safe, efficient, and scalable introduction of urban air mobility services.

CHAPTER 7

Environmental and Economic Impact

7.1 EMISSIONS REDUCTIONS POTENTIAL

Electric propulsion shifts decarbonization from a distant aspiration to a practical, staged plan for short-haul aviation. The headline potential—near-zero in-flight emissions for targeted routes—becomes durable only when paired with clean electricity, efficient operations, robust recycling, and transparent accounting. Hybrids, SAF, and air traffic optimization fill gaps while batteries advance, ensuring progress across a heterogeneous fleet.

Over the next decade, expect a mosaic of solutions: fully electric aircraft dominating short regional and urban missions, hybrids extending range with meaningful fuel cuts, airports becoming energy hubs with low-carbon microgrids, and operators competing not only on schedule and price but on verified grams of CO_2 per passenger-kilometer. The cumulative impact—route by route, fleet by fleet—can bend aviation's emissions curve down faster than physics alone would suggest, turning early, focused gains into a foundation for deeper sector-wide decarbonization.

7.1.1 Where Aviation Emissions Come From

Aviation's climate impact is dominated by CO_2 from burning jet fuel, with additional warming from non-CO_2 effects such as nitrogen oxides (NO_x) and contrails. Short-haul flights tend to be less fuel-efficient per passenger-kilometer than long-haul due to takeoff/climb penalties, lower load factors, and smaller airframes. Helicopters and small turboprops used for regional hops and urban missions also have high specific fuel consumption and emit NO_x in populated areas. This profile is precisely where electric propulsion can bite hardest: missions with high power during short segments, repetitive cycles, and operations near communities benefit disproportionally from high drivetrain efficiency and zero tailpipe emissions.

7.1.2 Accounting Boundaries: Tailpipe vs. Lifecycle

How we count emissions changes the answer. Tailpipe accounting is straightforward: fully electric aircraft have zero in-flight CO_2, NO_x, SO_x, and particulate emissions. Lifecycle or "well-to-wing" accounting includes electricity generation, battery manufacturing, and end-of-life, as well as upstream fuel extraction and refining for combustion aircraft. For electric aviation, two variables dominate lifecycle results: the grid carbon intensity where charging occurs (measured in g CO_2 per kWh) and the embodied emissions of the battery pack (kg CO_2 per kWh of capacity). Modern cell factories powered by cleaner grids are pushing pack manufacturing intensities toward roughly 40–80 kg CO_2 per kWh, and recycling can recover much of the material value, reducing future footprints. When amortized over thousands of cycles, the battery manufacturing overhead often contributes only a few grams CO_2 per passenger-kilometer, small compared with operational savings.

7.1.3 Physics Advantage of Electric Propulsion

Electric motors and inverters routinely deliver 85–95% efficiency from bus to shaft, compared with 25–40% for small turbines and piston engines at typical operating points. That efficiency dividend means less energy is needed for a given mission, even before considering the grid. For modern electric fixed-wing designs with high lift-to-drag ratios, energy intensities on the order of 0.12–0.25 kWh per passenger-kilometer are a plausible operating range depending on airframe, speed, and load factor. Multiply by a moderately clean grid (say 300 g CO_2/kWh) and operational emissions land in the 36–75 g CO_2/pkm range—well below many regional jets and turboprops, which often fall between 150 and 250 g CO2/pkm on short routes. Even on grids with higher intensity, electric's superior efficiency narrows the gap, and as grids decarbonize, the advantage widens.

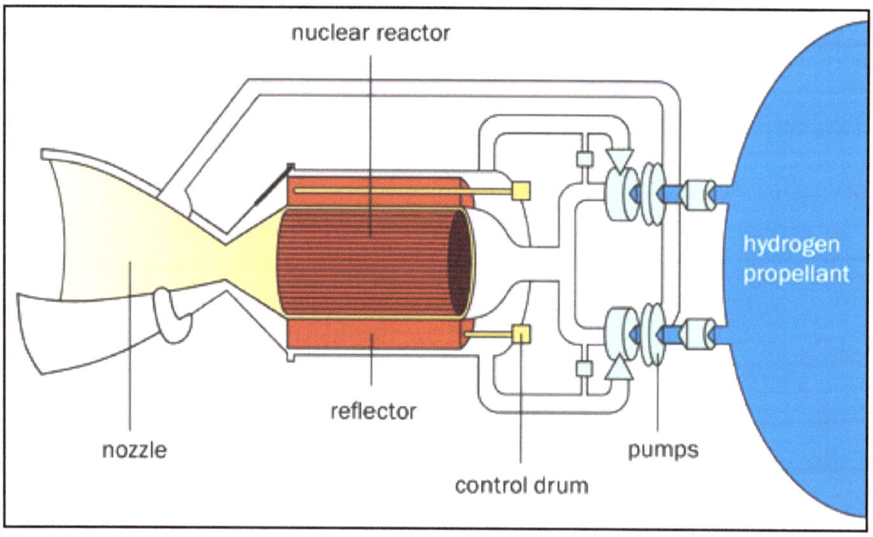

Fundamental Physics Benefits of Electric Propulsion

7.1.4 Sensitivity to Grid Carbon Intensity

Grid mix is the main swing factor for operational emissions of electric aircraft. In regions dominated by coal (e.g., 600–800 g CO_2/kWh), the emissions per passenger-kilometer rise proportionally; in hydro- or wind-rich regions (50–100 g CO_2/kWh), they plummet. Airports can mitigate this variability by contracting long-term renewable power purchase agreements, adding on-site solar and storage, and shifting charging to low-carbon hours. Energy storage at airports smooths peaks and aligns charge windows with cleaner power, while aircraft-side preconditioning reduces in-flight energy and thermal derates, further lowering total draw. Over a fleet and a year, these operational choices can swing emissions intensity by tens of percentage points, which is material at scale.

7.1.5 eVTOL vs. Fixed-Wing: Different Emissions Profiles

Vertical flight adds hover, which is power hungry and less aerodynamically efficient than wing-borne cruise. eVTOLs still benefit from electric efficiency, but their energy intensity depends heavily on mission design. Short urban hops with extended hover or loiter can consume 0.25–0.5 kWh per passenger-kilometer, while lift-plus-cruise configurations that minimize hover time and fly longer sectors at efficient speeds can approach or beat single-occupancy ground vehicles on emissions, particularly with clean power and high load factors. Acoustic and air-quality benefits close to cities remain compelling regardless. Fixed-wing electric aircraft, by contrast, are consistently efficient in cruise, making them the strongest near-term candidates for significant emissions reductions on 100–300 km regional routes.

7.1.6 Hybrid-Electric: Bridging Range and Reductions

Hybrid-electric designs—series, parallel, or turbo-electric—extend range and dispatch flexibility while capturing a portion of electric's efficiency benefits. A right-sized, steady-running turbogenerator can operate near its best efficiency island in cruise, while batteries handle climb and contingencies. The result is lower fuel burn and CO_2 than a conventional engine chasing throttle changes across the mission. Depending on architecture and route, fuel burn reductions of 10–30% are credible near-term targets, with higher gains on profiles where electric assist displaces the least efficient parts of the engine's operating map. Hybrids also eliminate most idling and enable electric taxi, trimming emissions on the ground where local air quality matters most. They cannot reach the zero tailpipe emissions of fully electric, but for longer regional legs and in grid-constrained regions, hybrids are a practical step-change.

7.1.7 Non-CO_2 Climate Effects and Air Quality

Electric propulsion removes in-flight NO_x and soot, the precursors for ozone formation and contrail ice nucleation. Contrails form primarily from water vapor and particulates in exhaust; with no combustion exhaust, electric aircraft essentially eliminate contrail-driven warming. There can be transient aerodynamic condensation in humid conditions, but its climate impact is small. On the ground and at low altitude, moving to electric flight (and electrified ground support equipment) reduces NO_x, SO_x, ultrafine particles, and volatile organics near populated areas—a public health benefit independent of climate accounting. Hybrids reduce, but do not eliminate, these pollutants; even so, shifting the most power-intensive segments to the battery lowers emissions spikes during takeoff and climb.

7.1.8 Lifecycle Impacts: Batteries, Materials, and End-of-Life

Batteries carry embodied emissions from mining, refining, and manufacturing. The industry is trending toward lower-intensity processes and renewable-powered gigafactories, which has already cut cell manufacturing footprints substantially. Recycling technologies that recover nickel, cobalt, lithium, and copper at high rates are maturing, lowering the need for virgin materials and avoiding waste. Structurally integrated packs can reduce total aircraft mass and thus operational energy, but must be designed for maintainability and safe end-of-life separation. When evaluated over service life—thousands of cycles under temperature-controlled, moderate-depth-of-discharge operation—battery manufacturing contributes a small fraction of total lifecycle emissions compared with combustion operations avoided.

7.1.9 Operations, Airports, and Systems Integration

How aircraft are flown and how airports supply energy materially affect emissions. Energy-aware flight management—optimal climb speeds, continuous descent, cruise altitudes tuned for minimal power draw—can shave 5–10% from energy use without hardware changes. At airports and vertiports, microgrids with storage, heat pumps, and renewables reduce the carbon intensity of charging and cut losses in thermal preconditioning. Standardized, high-voltage DC charging with active power factor correction reduces grid waste and enables higher utilization; shorter turn times lower the number of packs needed fleetwide, reducing the embodied footprint. Over time, coordinated scheduling between aircraft energy management systems and airport energy controllers will match real-time clean power availability to flights, turning data into emissions reductions.

7.1.10 Quantifying the Near-Term and Mid-Term Potential

In the near term, fully electric aircraft can credibly decarbonize specific niches: pilot training (a high-cycle, energy-light segment), island hops and short regional routes, and helicopter substitution for urban missions. Swapping combustion trainers for electric reduces both CO_2 and NO_x around airports; replacing short helicopter flights with eVTOLs slashes local emissions and noise. On a sector-wide basis, these niches may represent a single-digit percentage of total aviation CO_2 by 2030, yet they deliver outsized societal benefits and build the industrial and regulatory muscle needed for broader adoption. By the early-to-mid 2030s, as pack-level energy density climbs and certification experience accrues, electric fixed-wing aircraft should push deeper into 200–400 km routes, while hybrids trim fuel on 400–800 km legs. If paired with substantial grid decarbonization and SAF deployment on longer routes, the combined effect could plausibly reduce aviation's lifecycle emissions by 10–20% against a no-action baseline, with further gains as technology and infrastructure mature. The largest lever remains the grid: the cleaner the electricity, the closer operational emissions approach zero for electric segments.

7.1.11 High-Impact Levers to Maximize Emissions Reductions

- Prioritize routes where electric aircraft are most energy-efficient—fixed-wing regional legs and eVTOL corridors with minimal hover—to bank large CO_2 savings early.
- Contract low-carbon electricity for charging (PPAs, certificates) and add on-site solar plus storage to drive operational emissions toward zero regardless of grid volatility.
- Use energy-aware flight profiles and predictive AEMS to cut 5–10% from consumption through optimized climb, cruise, and continuous descent strategies.
- Precondition batteries at the stand to reduce in-flight thermal derates and power peaks, shrinking total energy use and improving cycle life.
- Electrify ground support equipment and taxi operations, and coordinate aircraft charging to avoid diesel generator use and cut scope 1 and 2 emissions at airports.
- Deploy hybrid-electric on longer regional routes to trim 10–30% fuel, while targeting full electrics where feasible; stack gains with SAF for residual emissions.
- Right-size airframes and avoid chronic oversizing; carrying excess structure or unused battery capacity raises per-seat energy and CO_2 intensity.
- Raise bus voltage and improve power electronics efficiency (SiC/GaN) to reduce electrical losses, allowing smaller packs for the same mission and lowering embodied emissions.
- Implement battery life-cycle management and recycling agreements to recover materials at high rates, cutting upstream mining impacts and reducing future pack footprints.

- Publish route-level, well-to-wing emissions and noise data to inform scheduling, earn community trust, and create feedback loops that drive continuous improvement.

7.2 LIFE CYCLE ASSESSMENT (LCA) OF ELECTRIC AVIATION

Life Cycle Assessment reveals electric aviation to be neither a silver bullet nor a marginal improvement, but a promising pathway whose benefits depend on careful choices across materials, energy, design, and operations. Electric propulsion delivers immediate gains in noise and local air quality and can substantially cut climate impacts when coupled with low-carbon electricity and thoughtful infrastructure. Yet batteries and composites carry significant embodied burdens that must be mitigated through cleaner manufacturing, better chemistries, circular design, and effective recycling. The industry's challenge is to align rapid technological progress with rigorous, transparent assessment so that early design decisions and policy frameworks lock in long-term advantages. With robust LCA guiding investment and regulation, electric aviation can mature into a cornerstone of a decarbonized, quieter, and more equitable air transport system.

7.2.1 Scope and Purpose of LCA in Aviation

The goal of applying LCA to electric aviation is to quantify environmental performance across the entire value chain and to inform better design, policy, and investment decisions. The functional unit—often expressed as passenger-kilometers or tonne-kilometers over the aircraft's useful life—anchors fair comparisons across technologies and operational profiles. Attributional LCA is commonly used to describe current state-of-play impacts, while consequential LCA helps explore system-level changes such as grid decarbonization or shifts in route networks. Properly defined system boundaries and data quality requirements are essential, because outcomes depend heavily on assumptions about flight cycles, load factors, duty cycles, and maintenance regimes.

7.2.2 System Boundaries and Functional Unit

Clear system boundaries prevent cherry-picking benefits. A robust LCA of electric aviation includes upstream mining and refining of battery minerals, manufacture of cells and packs, production of motors, power electronics, and lightweight airframes, transport of components, aircraft assembly, ground charging infrastructure, operations over a defined lifespan, maintenance and overhauls, and strategies for reuse, recycling, or disposal. Selecting an appropriate functional unit matters because short regional hops, helicopter-like urban air mobility missions, and longer thin-haul routes have distinct load factors, turnaround times, and energy intensities. Ensuring apples-to-apples comparisons against turboprops or regional jets requires harmonized assumptions about utilization and lifetime.

7.2.3 Inventory of Life Cycle Stages

The inventory phase quantifies inputs and outputs at each stage. Electric aircraft draw on lithium, nickel, cobalt, manganese, and graphite for batteries, copper and rare-earth elements in motors and generators, and carbon fiber, resins, and aluminum-lithium alloys for structures.

Mining and refining carry energy and water footprints as well as social risks, particularly for cobalt and nickel. Manufacturing is energy intensive, especially for carbon fiber precursors and autoclave curing, and for battery cell formation and conditioning. Operation ostensibly offers zero tailpipe emissions, but the electricity mix dictates indirect emissions and other impacts. End-of-life introduces complex pathways, as batteries demand dedicated recycling while composite structures challenge existing recovery methods.

7.2.4 Impact Categories and What they Reveal

Global warming potential often dominates discourse, yet LCA widens the lens to resource depletion, human and ecotoxicity, acidification, eutrophication, and photochemical smog formation, as well as water and land use. Electric aircraft tend to reduce operational greenhouse gases and eliminate in-flight nitrogen oxides, but may increase manufacturing-phase burdens due to battery and composite production. Toxicity risks cluster in upstream mining and in improper battery disposal. Acidification and eutrophication can arise from energy-intensive refining and from fossil-based electricity used in manufacturing. Noise, while not traditionally embedded in LCA metrics, is materially lower for electric propulsion and delivers tangible community benefits around airports and in urban corridors.

7.2.5 Electricity Mix and Operational Realities

Charging profiles and grid intensity determine whether electric aircraft translate design efficiency into climate benefits. In systems rich in wind, solar, hydro, or nuclear power, operational emissions can be dramatically lower than kerosene burning, especially on short regional routes where electrics excel. In coal-dominated grids, the advantage narrows or even vanishes when seen through a life cycle lens. Time-of-use charging, dedicated renewable power purchase agreements, on-site solar and storage at airports, and smart charging that aligns with surplus renewable generation can materially improve the LCA outcomes. Accurate modeling therefore requires temporal and regional granularity rather than a single average grid factor.

7.2.6 Batteries: Materials, Manufacturing, Use, and End-of-Life

Batteries are the heart of electric aviation and its most consequential LCA hotspot. Nickel-manganese-cobalt chemistries offer high energy density but rely on minerals with supply and social risks. Lithium iron phosphate reduces reliance on nickel and cobalt and offers long cycle life and safety advantages, though at a weight penalty that matters acutely in aviation. Solid-state, lithium-sulfur, and metal-air chemistries promise step-changes in energy density and safety, which could shift life cycle burdens by requiring fewer cells, enabling lighter airframes, and reducing operational energy.

Manufacturing remains energy hungry, with drying, calendaring, and formation cycles consuming substantial electricity; colocating gigafactories with clean grids markedly lowers embodied emissions. In use, battery degradation shapes replacement schedules and spares production, which ripple through the life cycle inventory. After aviation retirement, second-life applications in stationary storage can extend service value and amortize impacts before recycling. Hydrometallurgical, pyrometallurgical, and direct recycling pathways are advancing, yet collection logistics, disassembly, and safe transport are still maturing for large-format aviation packs.

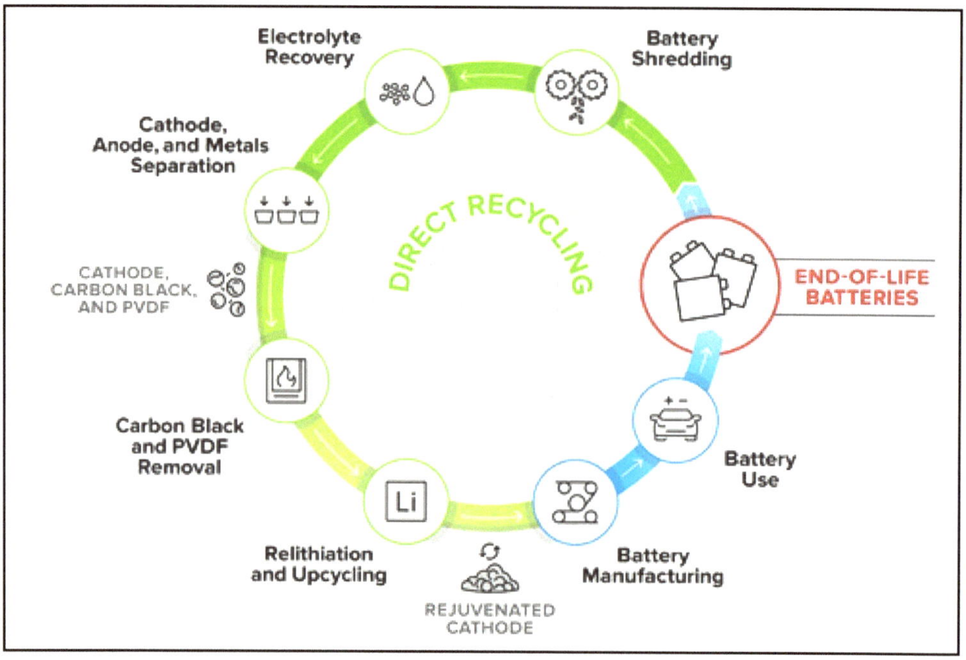

Batteries: Ingredients, Manufacturing Processes, Usage, and Retirement

7.2.7 Airframe, Propulsion, and Systems Manufacturing

Lightweight structures offset battery mass, but carbon fiber reinforced polymers carry high embodied energy due to precursor synthesis and high-temperature processing. Alternatives such as thermoplastic composites, bio-derived resins, and aluminum-lithium alloys can lower footprints while maintaining performance. Additive manufacturing of complex ducts, mounts, and heat exchangers minimizes scrap, shortens supply chains, and sometimes improves efficiency in service by reducing weight. Electric motors, inverters, and thermal management systems generally use fewer moving parts than turbine engines and can reduce maintenance burdens over the lifetime, yet rare-earth magnets, high-purity copper, and high-voltage semiconductors require careful sourcing and end-of-life recovery to avoid transferring impacts upstream.

7.2.8 Infrastructure and Ground Energy Ecosystems

Airports will require high-capacity charging, buffer storage, and smarter energy management to integrate fleets of electric aircraft without overtaxing local grids. Building this infrastructure entails steel, copper, concrete, transformers, switchgear, and battery energy storage systems, all of which add embodied impacts. Over time, infrastructure is amortized over thousands of flight cycles, so design choices that maximize utilization and align with renewable supplies become decisive. Urban air mobility vertiports, with their higher sortie rates and distributed locations, will depend even more on localized storage, microgrids, and automated charging to keep life cycle impacts low while maintaining operational reliability.

7.2.9 Comparative Insights with Conventional and Hybrid Platforms

When measured per passenger-kilometer, electric aircraft show their strongest advantages on short-haul regional routes where jet engines are most penalized by climb segments and poor seat factors. With low-carbon electricity, life cycle greenhouse gas reductions can be substantial even after accounting for battery and composite manufacturing. On grids reliant on fossil power, improvements shrink and may hinge on planned decarbonization trajectories over the aircraft's service life. Hybrid-electric architectures introduce additional complexity: they can reduce peak fuel burn during takeoff and climb and enable right-sized engines, but they add batteries, generators, and power electronics that increase manufacturing impacts and maintenance complexity. Their LCA performance depends on the balance between electrified mission segments, the cleanliness of electricity used to charge onboard storage, and any use of sustainable aviation fuels that carry their own upstream impacts.

7.2.10 Methodological Challenges and Uncertainties

Electric aviation is a moving target, and LCA must grapple with technological learning curves. Battery energy density, cycle life, and cobalt intensity are all changing quickly. Many datasets for motors, inverters, and large composite structures are proprietary and lack transparency, forcing analysts to rely on proxies that may under- or overestimate impacts. Choices between attributional and consequential approaches can alter conclusions, as can differing assumptions about aircraft lifetime, load factors, capacity growth, and maintenance practices. Geographical variation in electricity, water scarcity, and regulations around mining and waste complicate harmonization. End-of-life modeling often assumes recycling efficiencies that are not yet achieved at scale for aviation-grade components, creating optimistic or pessimistic biases depending on the scenario.

7.2.11 Strategies to Improve Life Cycle Performance

Design and policy levers exist across the life cycle to tilt outcomes in favor of electric aviation. Sourcing electricity from renewables through direct procurement or on-site generation cuts operational impacts and encourages grid upgrades that benefit communities more broadly. Engineering aircraft for disassembly, modular battery packs, and standardized interfaces streamlines second-life redeployment and materials recovery. Transitioning toward chemistries with fewer critical minerals reduces upstream risks, while investing in magnet and copper recycling addresses propulsion-system hotspots. Substituting lower-carbon materials and shifting composite manufacturing to clean-energy regions reduces embodied emissions. On the operational side, optimizing flight planning, thermal management, and charging strategies preserves battery life and improves energy efficiency, multiplying gains across thousands of cycles.

7.2.12 Policy, Certification, and Market Implications

Regulators and standards bodies can use LCA findings to encourage better technology choices without mandating a single pathway. Incentives for low-carbon manufacturing, clean charging infrastructure, and verified recycling can unlock private investment while setting guardrails against externalized harms. Certification frameworks that recognize whole-life emissions, not just tailpipe performance, push suppliers to disclose data and improve traceability for minerals and components. Public procurement and airport concession policies that reward verifiable life cycle improvements help early markets scale responsibly. Transparent communication about benefits and trade-offs builds trust with passengers and communities, countering greenwashing by grounding claims in standardized methods and third-party verification.

7.3 NOISE POLLUTION AND URBAN INTEGRATION

Noise pollution and urban integration represent two intertwined challenges at the forefront of the 21st-century aviation revolution. As electric and autonomous aircraft begin to populate urban skies, cities must evolve to balance technological innovation with environmental well-being. Effective management of noise involves not just quieter aircraft, but also thoughtful urban design, equitable planning, and community participation. By embedding acoustic considerations into every stage of development—from aircraft engineering to zoning laws—urban areas can embrace aerial mobility without compromising livability. The goal is not the complete elimination of noise, which is impossible in a dynamic city, but its intelligent management—transforming potential conflict into sustainable coexistence.

7.3.1 Nature and Sources of Urban Noise Pollution

Urban noise pollution arises from diverse sources including road traffic, railways, industrial operations, construction, and increasingly, aerial transportation. Aircraft noise, particularly during takeoff and landing, can penetrate deep into urban environments, disrupting sleep, communication, and mental well-being.

The acoustic footprint of traditional jet engines is well-documented, but even electric propulsion systems produce noticeable noise due to propeller rotation, airflow turbulence, and mechanical vibrations. In densely populated urban settings, where buildings amplify and reflect sound, the cumulative effect can create persistent background noise levels far above the recommended thresholds set by the World Health Organization (WHO).

7.3.2 Impact of Noise on Public Health and Quality of Life

Noise pollution is not merely an annoyance—it has profound physiological and psychological effects on urban populations. Continuous exposure to high noise levels can lead to hearing impairment, cardiovascular stress, sleep disruption, and reduced cognitive performance.

Children are particularly vulnerable, as noise interferes with learning and attention. Studies have shown correlations between chronic noise exposure and increased rates of hypertension and anxiety. Beyond individual health, noise pollution diminishes urban livability, influencing property values, public space usage, and community satisfaction. Addressing these impacts requires integrated policies that align public health objectives with transportation and urban development planning.

7.3.3 Role of Aviation in Urban Noise Profiles

Traditional aircraft operations have long been a major source of urban and suburban noise, especially near airports. The rapid emergence of eVTOL and drone-based delivery systems has brought renewed focus to aviation-related urban acoustics. Unlike conventional aircraft, eVTOLs are designed for short-distance travel and frequent operations at lower altitudes—closer to residential zones.

Although electric propulsion reduces engine noise, the distributed propeller systems and high-frequency tonal sounds can still cause discomfort if not properly managed. Therefore, urban planners, engineers, and regulatory bodies are developing Noise Abatement Flight Procedures (NAFPs) and flight corridor planning strategies to minimize exposure for city residents.

Table: Conventional Aircraft Noise vs. eVTOL Noise Characteristics

Parameter	Conventional Aircraft	eVTOL Aircraft
Primary Noise Source	Jet engine exhaust and combustion	Propeller and rotor turbulence
Frequency Range	Low to mid frequencies	Mid to high frequencies
Peak Noise Levels	120–140 dB near airports	60–80 dB in urban settings
Flight Altitude	1,000–35,000 feet	200–2,000 feet
Noise Management	Runway orientation and curfews	Flight path optimization and distributed propulsion design

7.3.4 Design Strategies for Noise Reduction in eVTOLs

Technological innovation plays a crucial role in reducing noise at the source. eVTOL manufacturers are employing distributed electric propulsion (DEP) systems, which use multiple smaller rotors operating at optimized speeds to distribute acoustic energy more evenly. Additionally, blade design optimization, such as reshaping rotor tips and improving aerodynamics, minimizes vortex interactions and tonal peaks. Advanced simulation tools now allow engineers to predict noise propagation patterns and adapt flight operations accordingly. Integration of active noise control technologies—similar to those in high-end headphones—is being explored to reduce cabin and external sound levels simultaneously. Collectively, these innovations contribute to making urban air transport quieter and more community-friendly.

7.3.5 Policy and Regulatory Frameworks for Urban Noise Management

Noise regulation in urban aviation is guided by national and international standards. Organizations such as the International Civil Aviation Organization (ICAO) and Federal Aviation Administration (FAA) have set noise certification standards that govern aircraft design and operations. Urban air mobility introduces new regulatory dimensions, as flight corridors will intersect with densely populated zones. Local governments are developing frameworks for acceptable noise thresholds, curfew hours, and community feedback mechanisms to ensure that aerial transport integrates responsibly. Data-driven policies, supported by noise monitoring networks and acoustic mapping, enable real-time assessments of environmental impact and inform adaptive management strategies for cities.

7.3.6 Urban Planning and Zoning for Acoustic Integration

Managing noise pollution is as much an urban planning issue as it is an engineering challenge. Cities can significantly mitigate acoustic impacts through careful zoning, land use planning, and infrastructure design. Establishing noise buffer zones around vertiports, using green barriers like trees and acoustic walls, and promoting mixed-use developments with transitional sound zones can reduce residential exposure. Rooftop vertiports on commercial or industrial structures, rather than near residential buildings, also help distribute acoustic load more effectively. Furthermore, urban design principles—including street orientation, building spacing, and façade insulation—can be leveraged to diffuse and absorb sound waves, improving overall acoustic quality in dense city environments.

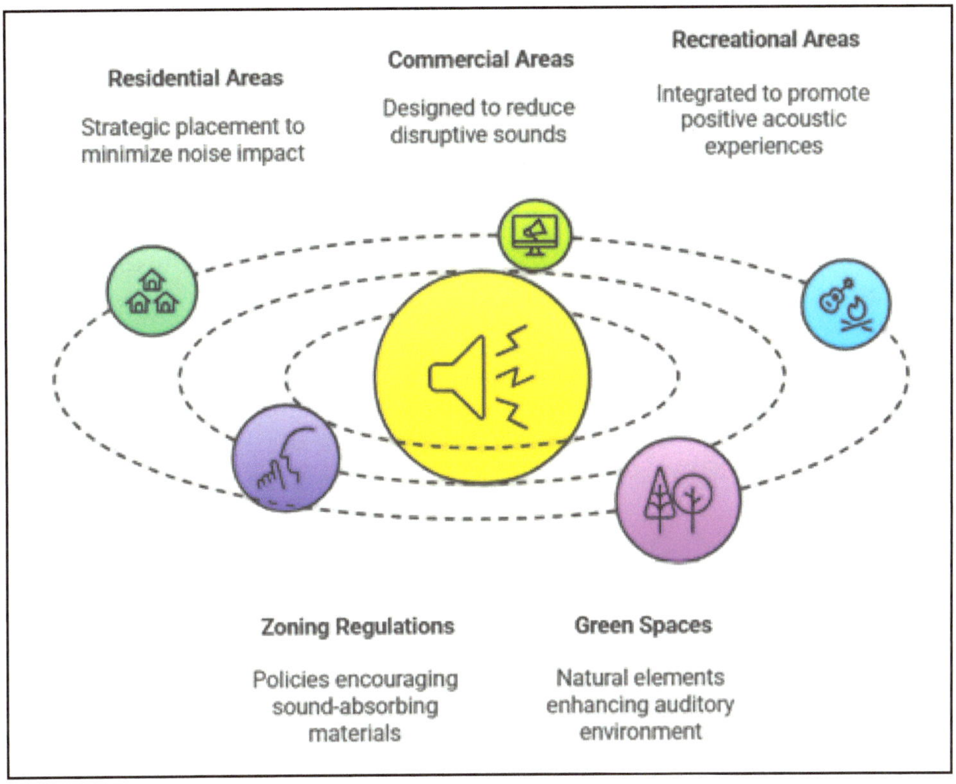

Creating Harmonious Urban Environments

7.3.7 Urban Noise and Integration

- Complexity in modeling acoustic propagation in dense urban environments
- Lack of standardized noise metrics for emerging aerial technologies
- Public resistance to perceived noise, even when objectively low
- Difficulty in enforcing flight path regulations in dynamic airspace
- Integration of vertiports within mixed-use cityscapes
- Balancing operational efficiency with community noise concerns
- Retrofitting existing buildings for noise resilience
- Coordination between aviation authorities and urban planners

- Need for real-time noise monitoring and public transparency
- Ensuring equitable noise distribution across urban districts

7.3.8 Technological and Operational Noise Mitigation Approaches

Noise mitigation extends beyond aircraft design to include operational strategies. Flight path optimization can reduce the concentration of noise over specific areas by distributing it more evenly. Automated air traffic management systems can adjust routes in real time based on population density and environmental data.

Low-altitude operations are minimized over sensitive zones such as hospitals and schools. On the ground, acoustic shielding structures around vertiports and maintenance facilities absorb takeoff and landing noise. Additionally, predictive acoustic modeling using artificial intelligence helps planners forecast the long-term effects of new flight corridors and vertiport locations, ensuring that urban integration is proactive rather than reactive.

7.3.9 Public Perception and Community Engagement

Community acceptance is one of the most critical determinants of successful urban air mobility integration. Even if noise levels are technically within regulatory limits, public discomfort or distrust can lead to opposition against new infrastructure. Transparent communication, data sharing, and participatory planning are essential tools for building trust.

Cities and aviation operators are increasingly conducting public acoustic demonstrations to help residents understand real-world sound levels of eVTOLs. Citizen feedback is also being incorporated into noise mapping tools and policy frameworks, ensuring that community perspectives directly influence urban air mobility planning. This participatory approach transforms noise management from a purely technical problem into a social contract between cities and their inhabitants.

7.3.10 Socioeconomic and Environmental Considerations

Noise pollution does not affect all communities equally—lower-income neighborhoods often face greater exposure due to their proximity to major transport corridors. Addressing this inequality is a crucial part of urban integration strategies. Equitable distribution of vertiports and flight routes ensures that no single district bears disproportionate acoustic burdens. From an environmental perspective, quieter eVTOLs contribute to reducing overall noise footprints, complementing their zero-emission benefits. Moreover, urban design interventions such as green roofs, vegetated façades, and sound-absorbing materials further enhance sustainability while mitigating noise. Integrating acoustic comfort into environmental and social policies promotes holistic urban development.

7.4 COST ANALYSIS: OPERATING VS. ACQUISITION COSTS

The effective management of both operating and acquisition costs represents a fundamental requirement for sustainable business success. Rather than focusing exclusively on minimizing one category of expenses, organizations must develop balanced strategies that optimize the relationship between these cost types to support overall business objectives. This balanced approach requires understanding the distinct characteristics, implications, and management requirements for each cost category while recognizing their interconnected nature.

Operating costs demand continuous attention to efficiency, productivity, and waste elimination to maintain competitiveness and profitability in the short term. Acquisition costs require strategic vision, careful analysis, and disciplined execution to build capabilities, capacity, and competitive advantage for long-term success. The most successful organizations integrate cost management into their strategic planning processes, aligning resource allocation decisions with business priorities, market opportunities, and competitive dynamics. They establish clear criteria for evaluating both operating and acquisition expenditures, implement robust systems for tracking and analyzing cost performance, and create accountability structures that empower managers to optimize costs within their areas of responsibility.

7.4.1 Defining Operating Costs: The Ongoing Business Expenses

Operating costs encompass all expenses required to maintain the day-to-day operations of a business. These costs are typically recurring in nature and are fully deducted as expenses in the accounting period in which they are incurred. The most common operating costs include employee salaries and wages, rent or lease payments for facilities, utilities such as electricity and water, raw materials and inventory, marketing and advertising expenses, insurance premiums, maintenance and repairs for equipment, administrative expenses, and taxes directly related to operational activities. These costs are recorded on the income statement and are subtracted from revenue to determine operating profit. The management of operating costs is crucial for maintaining profitability, as these expenses directly affect the company's bottom line. Businesses often focus on optimizing operational efficiency to reduce these costs without compromising quality or productivity. Regular monitoring and analysis of operating expenses help identify trends, pinpoint areas for potential savings, and maintain control over the financial aspects of business operations. Effective operating cost management requires continuous attention and adjustment as business conditions, market demands, and operational requirements evolve over time.

7.4.2 Acquisition Costs: The Initial Investments

Acquisition costs represent the capital investments made to obtain, improve, or maintain long-term assets that will generate value for the business over multiple accounting periods. These costs are not immediately expensed but are capitalized on the balance sheet and gradually depreciated or amortized over the useful life of the asset.

Common examples of acquisition costs include purchasing property, buildings, or equipment; acquiring vehicles or machinery; costs associated with developing or purchasing software systems; expenses related to business acquisitions or mergers; and significant improvements to existing assets that extend their useful life or increase their value. Unlike operating costs, acquisition costs represent strategic investments in the business's future capabilities and capacity. The decision to incur acquisition costs typically involves careful analysis of the expected return on investment, the asset's projected useful life, and the impact on the company's financial position. These investments often require substantial capital outlay and may involve financing through debt or equity. Proper management of acquisition costs includes thorough evaluation of alternatives, negotiation of favorable terms, and strategic timing of purchases to align with business cycles and financial capacity.

7.4.3 Differences between Operating and Acquisition Costs

The distinction between operating and acquisition costs extends beyond their basic definitions to encompass several critical financial and operational dimensions that influence business decision-making.

- **Accounting Treatment:** Operating costs are fully expensed in the period they are incurred, directly impacting the income statement and reducing taxable income for that period. Acquisition costs are capitalized as assets on the balance sheet and expensed over time through depreciation or amortization.
- **Tax Implications:** Operating expenses typically provide immediate tax deductions in the current fiscal year, while acquisition costs offer tax benefits gradually through depreciation schedules that may extend over several years.
- **Cash Flow Impact:** Operating costs represent recurring cash outflows that must be managed within regular business cycles, whereas acquisition costs often involve significant one-time cash outlays that may require special financing arrangements.
- **Strategic Importance:** Operating costs are associated with maintaining current operations and short-term efficiency, while acquisition costs represent strategic investments in future capabilities, capacity expansion, and long-term competitive advantage.

- **Decision-Making Timeframe:** Decisions regarding operating costs typically focus on short to medium-term horizons, often reviewed monthly or quarterly. Acquisition cost decisions involve long-term planning and evaluation, with consequences that may extend for years.
- **Flexibility and Reversibility:** Operating costs are generally more flexible and can often be adjusted relatively quickly in response to changing business conditions. Acquisition costs are typically more permanent commitments that are difficult to reverse without significant financial consequences.
- **Management Responsibility:** Operating cost management is usually decentralized across various departments and operational managers, while acquisition decisions are typically centralized at senior management levels with board oversight for major expenditures.
- **Financial Analysis Methods:** Operating costs are analyzed using metrics like operating ratios, efficiency measures, and variance analysis. Acquisition costs require more complex evaluation methods including return on investment calculations, payback periods, and net present value analysis.
- **Impact on Financial Statements:** Operating costs directly affect profitability measures on the income statement, while acquisition costs influence both the balance sheet through asset values and the income statement through subsequent depreciation expenses.
- **Risk Profile:** Operating costs generally involve lower risk as they support ongoing operations with predictable outcomes, while acquisition costs carry higher risk due to their substantial commitment and uncertain long-term returns.

7.4.4 Strategic Management of Operating Costs

Effective management of operating costs requires a systematic approach that balances cost control with operational effectiveness. The first step involves comprehensive tracking and categorization of all operating expenses to establish baseline measurements and identify spending patterns. Businesses should implement regular budget reviews and variance analyses to compare actual expenses against projected amounts and investigate significant deviations. Strategic operating cost management often includes negotiating favorable terms with suppliers, optimizing inventory levels to reduce carrying costs, implementing energy-efficient practices to lower utility expenses, and leveraging technology to automate processes and improve productivity. Many organizations adopt continuous improvement methodologies to identify and eliminate waste in operational processes, thereby reducing costs while maintaining or enhancing quality. Workforce management strategies, including appropriate staffing levels, training programs to improve efficiency, and performance-based compensation structures, can optimize labor costs which often represent a significant portion of operating expenses. Additionally, businesses should regularly evaluate outsourcing opportunities for non-core functions where external providers may offer superior cost efficiency. The most successful operating cost management strategies focus on creating sustainable cost structures that support business objectives rather than simply minimizing expenses, recognizing that strategic investments in certain operational areas can generate disproportionate returns through improved efficiency, quality, or customer satisfaction.

7.4.5 Strategic Management of Acquisition Costs

The management of acquisition costs requires a disciplined approach to capital investment decisions that aligns with long-term strategic objectives. Organizations should establish formal processes for evaluating potential acquisitions, including clearly defined criteria for project approval, standardized methods for calculating expected returns, and systematic assessment of risks associated with each investment. Effective acquisition cost management begins with thorough needs analysis to determine whether asset acquisition represents the optimal solution to business requirements, considering alternatives such as leasing, outsourcing, or process redesign. For approved acquisitions, businesses should implement competitive bidding processes, negotiate favorable payment terms, and consider timing strategies to maximize value and manage cash flow impact.

Post-acquisition, organizations must establish robust systems for tracking the performance of acquired assets against projected benefits, maintaining proper records for depreciation calculations, and implementing preventive maintenance programs to protect the asset's value and extend its useful life. Regular reviews of the organization's portfolio of capital assets help identify underperforming investments, opportunities for redeployment, or assets that may require upgrade or replacement. Strategic acquisition cost management also involves coordinating the timing of major capital expenditures with business cycles, cash flow availability, and financing options to minimize disruption to operational stability and financial health. By taking a strategic, disciplined approach to acquisition decisions and management, organizations can optimize their capital investments to support sustainable growth and competitive advantage.

7.4.6 Industry-Specific Considerations in Cost Analysis

The relative importance and management of operating versus acquisition costs vary significantly across different industries, reflecting unique business models, capital intensity, and operational characteristics. In manufacturing industries, acquisition costs for production machinery, facilities, and technology systems typically represent substantial investments that determine production capacity and capabilities, while operating costs include raw materials, labor, maintenance, and utilities that directly affect product cost and profitability.

Service industries generally have lower acquisition costs for physical assets but may significant operating costs related to professional labor, marketing, and facility operations. Technology companies often face substantial acquisition costs for research and development, intellectual property, and computing infrastructure, while their operating costs may be dominated by technical talent, marketing, and ongoing software maintenance. Retail businesses balance acquisition costs for store fixtures, inventory management systems, and retail locations against operating costs including inventory purchases, staffing, and marketing. Healthcare organizations manage significant acquisition costs for medical equipment and facilities alongside substantial operating costs for medical supplies, professional staff, and administrative functions. Transportation and logistics companies incur major acquisition costs for vehicles, aircraft, or shipping containers while managing operating costs including fuel, maintenance, and labor.

7.4.7 Impact of Technology on Cost Structures

Technological advancements continue to transform the relationship between operating and acquisition costs across virtually all industries. The emergence of cloud computing, software-as-a-service models, and other technology-enabled services has created new options for converting traditional acquisition costs into operating expenses. Businesses can now access sophisticated software, computing infrastructure, and business capabilities through subscription models that require minimal upfront investment but create ongoing operating expenses. This shift provides greater flexibility, reduces barriers to implementing advanced technologies, and transfers certain risks from the business to service providers.

Automation technologies including robotics, artificial intelligence, and process automation tools often require significant acquisition costs but can substantially reduce operating costs over time through improved efficiency, reduced labor requirements, and enhanced quality control. The Internet of Things enables predictive maintenance approaches that can optimize both acquisition decisions through extended asset life and operating costs through reduced downtime and maintenance expenses. Data analytics technologies represent acquisition costs for implementation but generate ongoing value through improved decision-making, operational insights, and identification of cost reduction opportunities. These technological trends require businesses to continually reassess their approach to both operating and acquisition costs, considering not only the financial implications but also strategic factors including flexibility, scalability, competitive positioning, and the ability to leverage emerging technologies for business advantage.

Technological Influence on Cost Frameworks

7.5 THE ROLE OF ELECTRIC AVIATION IN GREEN AVIATION GOALS

Electric aviation will not, by itself, deliver netzero aviation. But it is the tip of the spear for credible, incremental progress that compounds. Over the next decade, expect fully electric aircraft to dominate training, island and rural connectivity, and high-frequency urban missions where their efficiency and quietness align with real demand. Hybrids will extend range and reduce fuel burn on longer regional routes, bridging to fuel-cell and hydrogen options as those ecosystems mature. Airports will evolve into resilient energy hubs, turning megawatts and thermal capacity into reliable, lowcarbon operations. Regulators will refine noise and emissions metrics suited to urban contexts and give certification credit for validated digital models and datarich safety cases.

The strategic value is as much institutional as technological. Electric aviation forces the integration of energy, airspace, and community considerations into a single system—with standards, telemetry, and accountability at its core. That integration is the blueprint for the sector's broader transition: electrify where possible, decarbonize energy supply, optimize operations, recycle materials, and communicate openly. If pursued with rigor and humility, the role of electric aviation in green aviation goals can be both immediate and catalytic: cleaner skies, quieter neighborhoods, and a faster glidepath to truly sustainable flight.

7.5.1 Green Aviation Goals and where Electric Fits

Global goals are clear: the International Civil Aviation Organization (ICAO) has adopted a longterm aspirational goal of netzero CO_2 by 2050, complemented by marketbased measures like CORSIA for international flights and regional policies that price carbon and mandate sustainable fuels. National and regional initiatives—such as ReFuelEU Aviation in Europe, SAF incentives in the United States, and emerging AAM frameworks—translate these goals into practical levers.

Within this mosaic, electric aviation plays three roles. First, it directly decarbonizes shorthaul segments where batteries can supply mission energy today. Second, it eliminates local pollutants and reduces noise, advancing public health and social license. Third, it accelerates the sector's electrification learning curve—power electronics, highvoltage safety, thermal management, and energy-aware operations—that benefits hybrids, hydrogen-electric, and even conventional aircraft via moreelectric subsystems.

7.5.2 Efficiency Dividend: Doing More with Less Energy

Electric propulsion converts a much larger fraction of input energy into useful thrust than combustion engines. Motors and inverters typically achieve 85–95% efficiency from bus to shaft, compared with 25–40% for small turbines and piston engines in common operating regimes. That physics advantage shrinks the energy needed for a given mission. Fixedwing electric aircraft with efficient aerodynamics can operate in the neighborhood of roughly 0.12–0.25 kWh per passengerkilometer depending on load factors and speed. Even before accounting for cleaner electricity, this efficiency lowers the emissions floor. As grids decarbonize and airports procure renewable power, operational emissions drop proportionally—turning a technology advantage into a climate outcome.

7.5.3 Near-Term Emissions Reductions where it Matters Most

Shorthaul aviation is disproportionately carbon and noise intensive per passengerkilometer because takeoff and climb dominate energy use and load factors are often modest. This is precisely where electric propulsion shines. Training flights, island hops, and 100–300 km regional sectors can be served by fixedwing electrics under today's battery constraints.

In urban and suburban corridors, electric VTOL (eVTOL) aircraft can substitute for helicopter missions—medical logistics, infrastructure inspection, pointtopoint shuttles—at a fraction of the noise and with zero tailpipe emissions. For longer regional routes, hybridelectric designs extend range while delivering doubledigit fuel savings by letting a rightsized turbogenerator run near its bestefficiency point and reserving batteries for climb and contingencies. Each of these use cases unlocks tangible, bankable emissions cuts in the 2020s, rather than decades from now.

7.5.4 Beyond CO_2: Noise, Air Quality, and Community Health

Green aviation goals extend beyond climate. Airportadjacent communities bear the brunt of NOx, ultrafine particles, and high peak noise events that affect sleep, learning, and cardiovascular health. Electric aircraft eliminate inflight NOx and soot and shift the acoustic signature from combustion and jet exhaust to controllable aerodynamic sources.

Larger, slower rotors and propellers, distributed propulsion, and precise RPM control reshape the soundscape, reducing both perceived loudness and the sharp tonal spikes associated with helicopters. Quieter operations expand permissible windows and make vertiports viable in urban settings. Electrifying ground support equipment and taxi operations further cuts emissions and noise at the ramp, improving air quality for workers and nearby residents.

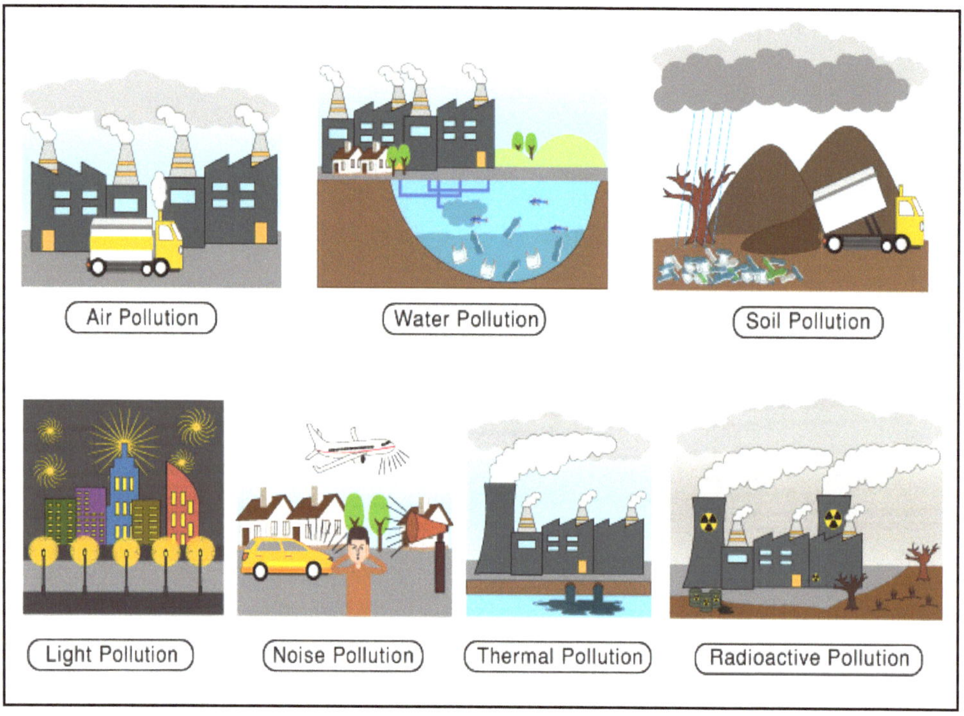

Beyond Carbon: Noise, Air Quality, and the Health of Communities

7.5.5 Infrastructure as an Environmental Force Multiplier

Airports and vertiports are becoming energy hubs. Highvoltage DC charging, onsite solar, battery energy storage systems, and smart thermal plants work together as microgrids that deliver reliable power while minimizing carbon intensity. By charging aircraft during lowcarbon hours, preconditioning batteries to avoid inflight derates, and shaving peak demand to reduce losses, airports convert infrastructure decisions into direct emissions savings. This same infrastructure decarbonizes the wider airport ecosystem—ground vehicles, buildings, and thermal systems—aligning aviation's green goals with city and utility transition plans. Importantly, interoperable standards and modular designs help airports scale without stranding assets as aircraft capabilities evolve.

7.5.6 Lifecycle Sustainability: Batteries, Materials, and Recycling

Responsible electrification accounts for the full lifecycle. Batteries embody emissions from mining and manufacturing; those impacts are falling as factories adopt renewable power and efficient processes. Aviationgrade packs can spread embodied carbon over thousands of tightly controlled cycles, making the perflight overhead small relative to operational savings. Design for disassembly and secondlife use in stationary storage preserves material value, while modern recycling recovers nickel, cobalt, lithium, and copper at high rates, reducing the need for virgin extraction. Material choices also matter: chemistries with lower critical mineral content, structural packs that eliminate redundant mass, and highrecovery aluminum and composite pathways all push electric aviation closer to circularity, a key pillar of comprehensive green goals.

7.5.7 Complementarity with SAF and Hydrogen

Electric aircraft do not compete with sustainable aviation fuels (SAF) and hydrogen—they complement them. SAF can decarbonize longer-haul flights that batteries cannot reach in the near term, and its production ramp benefits from demand signals where electrics are not yet viable. Hydrogen, particularly in fuel-cell electric architectures, holds promise for midrange missions as storage and systems mature.

Meanwhile, electric aviation reduces overall sector demand for liquid fuels, reserving limited SAF supply for where it matters most, and builds core capabilities—highvoltage safety, thermal integration, power management—that hydrogenelectric will also need. A portfolio approach aligns with green aviation goals: electrify what you can now, hybridize next, deploy SAF where needed, and keep pushing the technology frontier.

7.5.8 Equity, Access, and the Social License to Scale

Green goals are social goals. Electric aircraft enable new patterns of mobility that can be both cleaner and more inclusive: serving thin regional routes with smaller, quieter aircraft; linking rural clinics to urban hospitals with reliable medical logistics; and providing affordable training pathways for a more diverse pilot pipeline. Lower operating costs and noise footprints can reopen underused airports and create economic activity without imposing environmental burdens on neighbors. Transparent measurement—publishing routelevel emissions and noise data—and responsive operations that adapt flight paths and schedules based on community feedback are essential to earning and keeping social license as networks expand.

7.5.9 Measurement, Certification, and Proof

To count, improvements must be measured. Welltowing accounting that includes electricity generation, charging losses, thermal preconditioning, and battery lifecycle provides a credible basis for emissions claims. Noise must be characterized with eventbased metrics that reflect urban realities, not only traditional airport procedures. Certification pathways for electric aircraft and ground systems embed safety and environmental performance into airworthiness and operational approvals, ensuring that climate and community benefits are delivered in practice. Digital twins and fleet telemetry close the loop, turning operational data into continuous improvement—more accurate energy prediction, quieter profiles, smarter charging—and transparent reporting to regulators, customers, and communities.

7.5.10 Actions to Maximize Electric Aviation's Contribution to Green Goals

- Target missions where electric aircraft are most efficient—fixedwing regional routes and eVTOL corridors with minimal hover—to bank large, nearterm emissions and noise gains.
- Contract lowcarbon electricity for charging via PPAs or certificates, and build onsite solar plus storage to decouple operations from grid volatility and cut lifecycle emissions.
- Standardize interoperable, highvoltage DC charging and secure communications so mixed fleets can share infrastructure, improving utilization and reducing embodied impact.
- Integrate airport microgrids with predictive scheduling and aircraft energy management systems to charge during lowcarbon windows and precondition packs efficiently.
- Electrify ground support equipment and taxi operations, and coordinate ramp energy to eliminate diesel generators and reduce local pollutants at the source.
- Design for lifecycle: choose chemistries with robust recycling pathways, plan secondlife storage for retired packs, and specify suppliers with lowcarbon manufacturing.
- Use energyaware flight profiles—optimized climb, cruise, and continuous descent—and enforce realistic reserves with validated stateofenergy models to cut consumption without compromising safety.

- Deploy hybridelectric on longer regional legs and stack with SAF where needed; treat electrification, hybridization, and SAF as complementary tools, not substitutes.
- Publish routelevel, welltowing emissions and noise maps, engage communities in procedure design, and tie operational approvals to transparent performance.
- Build workforce skills in highvoltage safety, power electronics, and thermal systems; a trained workforce keeps systems efficient, safe, and available, amplifying environmental benefits.

CHAPTER 8

Hybrid-Electric and Alternative Power Models

8.1 PARALLEL AND SERIES HYBRID ARCHITECTURES

Parallel and series hybrid architectures represent two coherent philosophies for merging mechanical and electrical propulsion. Parallel hybrids thrive on versatility and highway efficiency by preserving a direct mechanical path and using electric torque as an enhancer. Series hybrids excel in urban and stop-go duty by decoupling the engine from the wheels, running it where it is cleanest and most efficient, and delivering a consistently electric driving feel.

The better architecture is not absolute but conditional on use case, packaging, and market context, and the best solutions often blend elements of both. With maturing batteries, smarter control, and cleaner grids, each architecture can deliver compelling efficiency, emissions reductions, and drivability improvements. The enduring lesson is that architecture, component sizing, and software must be designed as a system, matched to real-world duty cycles and supported by manufacturing and recycling choices that sustain their gains beyond the test cycle.

8.1.1 Fundamentals of Hybridization

At its core, hybridization pairs an internal combustion engine with at least one electric motor, an energy storage device, and power electronics that convert and control energy flows. The engine is dense in energy and fast to refuel, while the motor offers instantaneous torque, regenerative braking, and the potential for zero-emission operation during short trips. Batteries, typically lithium-ion with chemistries tuned for power or energy, buffer transients, recapture braking energy, and enable engine-off operation. Inverters and DC/DC converters orchestrate these flows, while software aligns them with driver demand, emissions constraints, and battery state-of-charge goals. How and where these elements connect mechanically yields the distinction between parallel and series hybrids.

8.1.2 Parallel Hybrid Architecture

Parallel hybrids connect the engine and electric motor to the same driveline, allowing both machines to deliver torque directly to the wheels. This configuration preserves the conventional mechanical path from engine to transmission to axle, supplementing it with electric torque that can assist during launches, fill torque during gear shifts, or propel the vehicle alone under light loads. The result is a flexible and efficient system that can capitalize on engine strengths at steady speeds while exploiting the motor's responsiveness for transient events.

Within the parallel family, component placement along the driveline shapes performance and packaging. A belt-integrated starter generator mounted on the front of the engine can enable mild hybrid features like stop-start, coasting, and limited torque assist without major transmission changes, making it attractive for cost-sensitive applications.

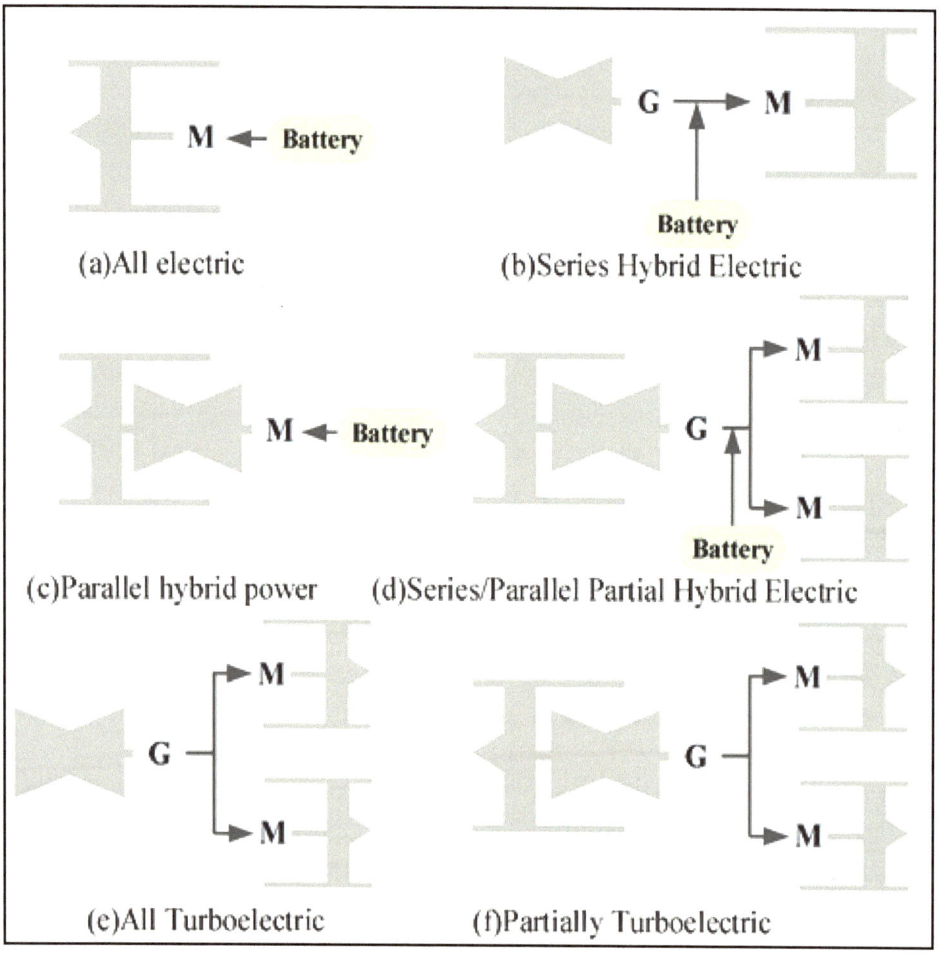

Parallel Hybrid Powertrain Architecture

Locating the motor on the crankshaft ahead of a clutch integrates deeper electric assist but may drag the engine unless a disconnect is used. Placing the motor between engine and transmission with a dedicated clutch enables true engine decoupling, allowing modest electric-only driving without spinning engine internals and improving regenerative braking efficiency. Motors mounted at the output shaft or on the axle, including through an e-axle on the rear, can add all-wheel-drive capability and increase regen potential, though coordination with the transmission becomes more complex. These variations are frequently described in industry shorthand based on motor position relative to the engine and gearbox, and they reflect trade-offs among costs, efficiency, and drivability.

Operationally, parallel hybrids employ several modes. In engine-only mode, the system behaves like a conventional powertrain, often with the engine calibrated for efficient high-load operation due to the motor's ability to handle transients. In blended hybrid mode, both sources contribute torque, smoothing shifts and reducing engine excursions into inefficient regions. In electric-only mode, feasible at low speeds and loads depending on motor power and battery capacity, the vehicle can operate quietly and with zero tailpipe emissions. Regenerative braking captures kinetic energy via the motor during deceleration, converting it to stored electrical energy that offsets future fuel consumption. Control software decides which mode to use, typically based on driver demand, battery state-of-charge, catalyst temperature, and anticipated route conditions.

The strengths of parallel architectures are rooted in their versatility and efficiency on diverse drive cycles. Because mechanical losses are reduced at steady cruise and the engine can operate near its efficiency islands, highway fuel economy is strong. The motor's ability to supplement torque enables engine downsizing and downspeeding, improving thermal efficiency without sacrificing performance. Packaging can be friendly to existing vehicle platforms by modifying or replacing transmissions with dedicated hybrid units that integrate electric machines and clutches. The primary compromises involve complexity in clutches and gearsets, the need for sophisticated shift and torque blending control, and limited pure-electric range unless motor and battery sizes are increased. When constrained by cost and space, many parallel hybrids prioritize power assist and regenerative braking over extended EV operation, striking a pragmatic balance for mixed-use driving.

8.1.3 Series Hybrid Architecture

Series hybrids break the mechanical link between the engine and the wheels, assigning the engine exclusively to a generator that produces electricity for the traction motor and battery. In this layout, the electric motor is the sole provider of wheel torque, and the engine turns on only to maintain battery charge or provide sustained power through the generator. This architecture transforms the vehicle into an electric drive system with an onboard power plant, often described as a range extender when the engine is small and primarily a backup. By removing the mechanical coupling to the wheels, the engine in a series hybrid can operate at a narrow set of speeds and loads where brake-specific fuel consumption is lowest, reducing pumping and friction losses and simplifying emissions control. Transients are largely absorbed by the battery and motor, enabling smooth acceleration and consistent engine operation that is easier on aftertreatment systems. Packaging can be more flexible because the engine-generator set does not need to align with the driveline, which is helpful in applications like buses, delivery trucks, railcars, or marine craft where modularity and accessibility matter.

However, series hybrids necessarily convert engine output from mechanical to electrical and back to mechanical at the wheels, incurring conversion losses in the generator, power electronics, battery, and motor. At sustained high-speed highway conditions where the engine in a conventional or parallel car could directly drive the wheels efficiently, the series architecture may trail on efficiency due to compounded conversion losses. To compensate, the motor must be sized for peak traction demand because it is always the final drive source, which raises cost and may increase copper usage and thermal management demands. These trade-offs explain why series hybrids excel in stop-and-go urban cycles with frequent braking opportunities and modest cruising, and why they are prevalent in buses and urban logistics where duty cycles amplify their strengths.

8.1.4 Comparative Performance and Use Cases

Drive cycle sensitivity is the defining difference between parallel and series hybrids. In urban conditions with recurrent starts and stops, a series hybrid can keep the engine in its sweet spot or off altogether and harvest significant regenerative energy, delivering strong real-world efficiency and low local emissions. In highway-dominant conditions with steady high loads, a parallel hybrid's ability to bypass electrical conversion and couple engine torque directly to the road becomes an advantage. Battery capacity magnifies these differences. Plug-in variants of either architecture can deliver substantial all-electric miles on daily commutes, after which their hybrid behavior governs longer trips. A plug-in series hybrid, often marketed as an extended-range electric vehicle, offers an electric driving experience with the engine serving as a quiet, steady generator for long journeys. A plug-in parallel hybrid can toggle between EV operation and efficient blended modes, particularly if equipped with a dedicated hybrid transmission that offers low-loss mechanical paths at cruise.

Segment-specific considerations also influence architectural fit. Heavy-duty transit buses, refuse trucks, and port equipment operating on short routes benefit from series designs that accommodate frequent idling, high regen fractions, and simplified drivetrains. Passenger cars and crossovers that split time between city and highway often favor parallel layouts for better long-distance efficiency and wide adoption compatibility with existing platforms. Niche formats like range-extended urban cars prioritize the electric driving feel and low NVH that series hybrids offer, leveraging small, acoustically isolated generators. Rail and marine have long exploited series principles, pairing diesel engines with generators feeding traction motors, where torque control and modularity outweigh conversion losses.

8.1.5 Control, Energy Management, and Software

Both architectures depend on sophisticated supervisory control to convert hardware potential into real-world gains. Rule-based strategies manage engine on-off decisions, set battery state-of-charge windows, and coordinate thermal demands, while optimization techniques like equivalent consumption minimization and model predictive control adjust torque splits and generator scheduling in real time to minimize fuel and emissions. Predictive information from navigation, traffic, and topography helps precharge before climbs, exploit downhill regen, and defer engine operation away from sensitive zones. In parallel hybrids, clutch coordination and torque shaping ensure seamless shifts and brake feel that blends friction and regeneration transparently. In series hybrids, generator speed maps, noise abatement strategies, and battery temperature limits govern when and how the engine contributes, balancing efficiency with cabin comfort.

8.1.6 Component Sizing, Materials, and Thermal Management

Sizing choices flow from architecture. Parallel systems can get by with lower-power motors because the engine can always share traction duty, while series systems require motors sized to peak road loads and generators scaled to sustained power needs. Batteries in series hybrids often emphasize power capability and thermal resilience to handle frequent cycling, whereas parallel systems can sometimes prioritize energy modestly if pure EV range is not a priority. Thermal management spans battery, motor, inverter, generator, and engine, with integrated cooling loops increasingly common to reduce mass and improve heat rejection. Advances in wide-bandgap semiconductors like silicon carbide lower inverter losses and shrink cooling requirements, while motor designs using permanent magnets or magnet-free topologies trade off efficiency, rare-earth use, and cost. Packaging strategies favor compact e-axles and dedicated hybrid transmissions in parallel layouts, and modular generator packs in series designs that simplify service and upgrades.

8.1.7 Emissions, NVH, and Drivability

Hybrids aim to reduce both carbon and pollutant emissions across diverse conditions. Parallel systems must manage cold-start emissions by warming catalysts efficiently and avoiding engine idling when aftertreatment is not lit. Series systems can run the engine at high efficiency and catalyst-friendly conditions, often reducing NOx and particulates in city use, and can delay or batch engine operation until components are at temperature. From an NVH perspective, parallel hybrids must mask engine starts and blend torque to feel seamless, whereas series hybrids need strategies to avoid abrupt generator engagement at odds with driver expectations; careful calibration of sound, vibration isolation, and engine operating points can turn the generator into background noise. Drivability benefits accrue in both architectures through instant electric torque, torque fill during shifts, and smoother low-speed control in traffic.

8.1.8 Plug-In Dimension and Grid Interactions

Adding plug-in capability reframes the efficiency calculus by enabling grid energy to displace fuel. A plug-in parallel hybrid can cover daily trips electrically and reserve the engine for longer excursions, while maintaining efficient mechanical drive at highway speeds. A plug-in series hybrid accentuates the EV experience by operating as a pure electric until its buffer is depleted, after which the generator sustains operation. Grid carbon intensity and charging behavior determine lifecycle emissions outcomes, making at-home off-peak or renewable-sourced charging valuable. Vehicle-to-load and vehicle-to-grid features are starting to appear, but they impose additional cycle aging on batteries that control strategies must account for, particularly in series systems that already cycle frequently.

8.1.9 Costs, Reliability, and Serviceability

Cost structures differ with architecture. Parallel systems can reuse existing transmissions and leverage smaller motors, but they add complexity with clutches and integrated gearsets that require precise control and tuning. Series systems simplify the mechanical driveline but demand larger motors, robust generators, and high-power electronics, shifting cost and reliability questions into electrical domains. Service patterns reflect these choices. Brake life typically improves in both architectures thanks to regeneration, while engine wear can decrease due to fewer transients and optimized operation. Battery longevity depends on thermal control, depth-of-discharge patterns, and charge rates, with series hybrids often benefiting from narrower state-of-charge windows but incurring more cycling.

8.1.10 Technology Trajectories and Convergence

As batteries improve in energy density and cost, the relative importance of hybridization changes, yet the architectures also evolve. Dedicated hybrid transmissions for parallel systems reduce losses and expand electric-mode capability without sacrificing cruising efficiency. Series hybrids are experimenting with variable-speed generators, advanced noise management, and alternative prime movers like microturbines or hydrogen fuel cells to eliminate combustion entirely. Software continues to be the differentiator, with connected energy management that anticipates routes and grid conditions. Some manufacturers adopt blended power-split approaches that merge series and parallel benefits through planetary gearsets, a reminder that the design space is continuous rather than binary. High-voltage systems, silicon carbide electronics, and compact e-axles are shrinking the efficiency gap between modes and improving packaging freedom, while sustainability pressures push magnet recycling, reduced critical material use, and lifecycle-focused design.

8.2 RANGE EXTENDERS AND TURBOELECTRIC CONCEPTS

Range extenders and turboelectric propulsion systems are critical innovations in the evolution of electric aviation. They merge the reliability and energy density of combustion technologies with the efficiency and environmental benefits of electric propulsion. By extending range and enabling more flexible aircraft architectures, these systems offer a practical pathway toward cleaner air travel.

While technical and regulatory challenges remain, ongoing advancements in materials, control systems, and energy storage promise to make hybrid and turboelectric aircraft viable within the next decade. Ultimately, these technologies represent not just a bridge but a cornerstone of sustainable aviation, ensuring that the industry moves steadily toward a zero-emission future without sacrificing performance, safety, or economic viability.

8.2.1 Need for Range Extension in Electric Aviation

Electric aircraft rely solely on battery-stored energy to power their motors. However, modern lithium-ion batteries store only a fraction of the energy per kilogram compared to aviation fuel. This disparity creates a fundamental limitation for range and payload capacity. Range extenders help address this issue by providing an auxiliary source of electrical energy during flight, effectively recharging batteries or directly powering electric motors. Such systems make it possible for electric aircraft to undertake regional and mid-range missions without fully depleting battery reserves. Range extension technologies are essential for reducing operational constraints and making electric aviation practical in commercial contexts.

8.2.2 Concept and Design of Range Extender Systems

A range extender is a secondary power unit designed to generate electricity to support propulsion during flight. It is not meant to directly drive the aircraft mechanically but rather to supplement the main battery system. Several types of range extenders have been explored in aviation, including small piston engines coupled with generators, micro gas turbines, and hydrogen fuel cells. The design aims to achieve high efficiency, low emissions, and minimal added weight. The system operates under optimal conditions to provide steady power output, reducing the load on the main batteries. Power management systems play a crucial role in distributing energy between the extender, batteries, and propulsion units while maintaining overall flight balance and safety.

Table: Fundamentals and Configuration of Range Extender Systems

Range Extender Type	Energy Source	Efficiency	Emissions	Technology Readiness
Piston Engine Generator	Aviation fuel	Moderate	Medium	High
Micro Gas Turbine	Jet fuel, biofuel	High	Low–Medium	Medium
Hydrogen Fuel Cell	Hydrogen	Very High	Zero (if green hydrogen)	Low–Medium
Solid Oxide Fuel Cell	Hydrogen or natural gas	High	Low	Experimental

8.2.3 Turboelectric Propulsion: A Step Toward Hybrid Efficiency

Turboelectric propulsion integrates a gas turbine with electric generators and motors. The turbine drives a generator, which produces electricity that powers distributed electric propulsion units. This setup separates the energy generation from thrust production, allowing more flexible aircraft design. Turboelectric systems are particularly suited for regional or medium-haul aircraft, offering efficiency improvements through optimized turbine operation and distributed power delivery. Unlike conventional aircraft engines that mechanically drive propellers or fans, turboelectric systems convert mechanical energy into electrical energy and then distribute it to multiple motors. This architecture enables better weight distribution, improved aerodynamic efficiency, and potentially lower noise levels.

8.2.4 Benefits of Range Extenders and Turboelectric Systems

Range extenders and turboelectric systems provide numerous benefits for modern aviation. They allow electric aircraft to achieve significantly longer ranges while maintaining lower emissions compared to conventional propulsion. They also serve as a transitional technology, enabling the industry to advance toward sustainability without waiting for revolutionary breakthroughs in battery density.By operating turbines or engines at their most efficient points, these systems reduce overall fuel consumption. Additionally, the flexibility of electric power distribution allows designers to explore novel aircraft configurations, such as blended-wing bodies and distributed propulsion layouts, which improve lift-to-drag ratios and further reduce energy requirements.

8.2.5 Technological Challenges and Integration Barriers

Despite their advantages, hybrid and turboelectric systems introduce considerable complexity. Managing multiple power sources requires sophisticated control systems to ensure stability, efficiency, and safety. Weight and thermal management are also major concerns. The addition of turbines, generators, and cooling systems increases overall mass, which can offset efficiency gains. Power electronics represent another technical hurdle. High-voltage systems must operate with minimal losses, necessitating the use of advanced semiconductor materials like silicon carbide (SiC) and gallium nitride (GaN). Certification standards for hybrid-electric aircraft are still evolving, demanding rigorous testing of electrical insulation, redundancy, and fire protection systems before widespread adoption can occur.

8.2.6 Environmental Impacts and Fuel Considerations

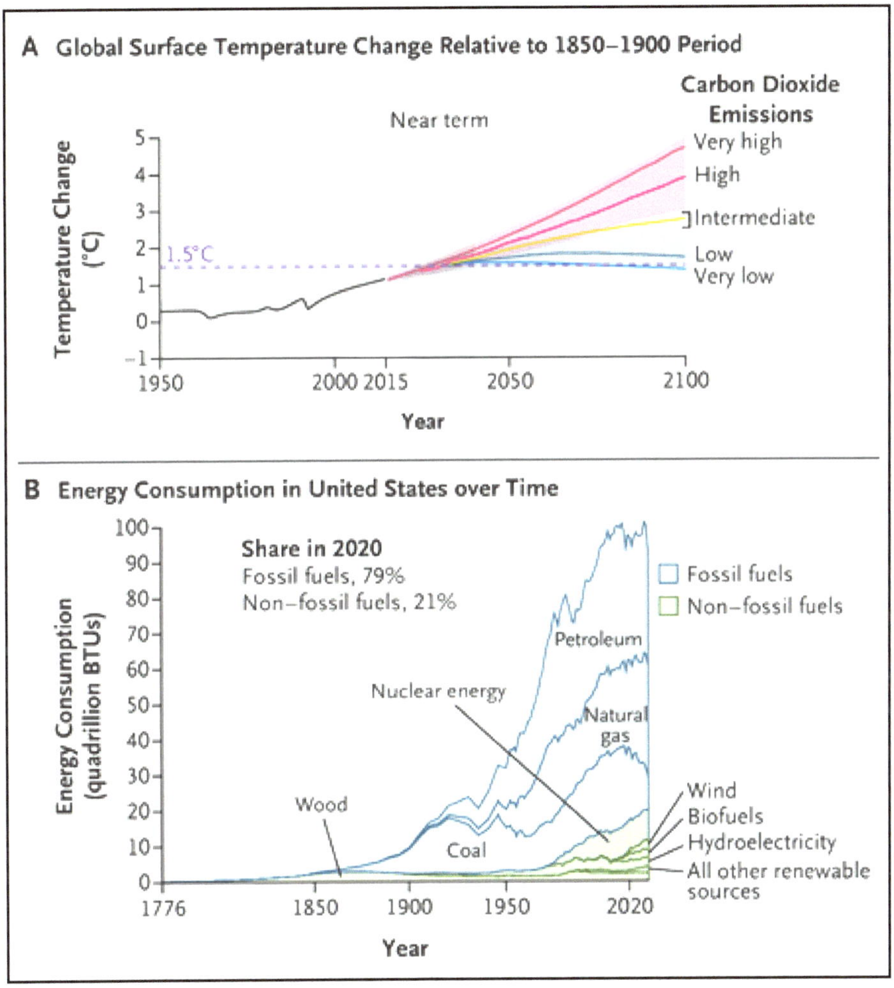

Impact on the Environment and Fuel Considerations

Hybrid-electric aircraft with range extenders can significantly reduce carbon dioxide emissions compared to conventional aircraft. The extent of reduction depends on the type of range extender and fuel used. When coupled with sustainable aviation fuels or hydrogen, emissions can be minimized to near-zero levels. Hydrogen-based systems, especially fuel cells, are particularly promising due to their high efficiency and water-vapor-only exhaust. Nevertheless, these systems are not entirely free from environmental challenges. The production of hydrogen and sustainable fuels requires substantial energy input, and their availability remains limited. Infrastructure for hydrogen storage, transportation, and refueling is still in its infancy. Therefore, hybrid-electric systems represent a near-term step toward decarbonization, balancing current technical feasibility with long-term sustainability goals.

8.2.7 Applications and Emerging Projects

Several ongoing research programs and prototypes demonstrate the potential of range extenders and turboelectric propulsion. Airbus's E-Fan X project, in collaboration with Rolls-Royce and Siemens, tested a hybrid-electric system where a turbine powered a generator to drive one of the aircraft's electric fans. Although discontinued in 2020, it provided valuable insights into system integration and high-voltage power management. NASA has also been actively researching turboelectric propulsion concepts, notably through the X-57 Maxwell and subsequent hybrid initiatives. Smaller companies such as Ampaire and VoltAero have been developing hybrid-electric aircraft designed for regional and commuter flights. These aircraft use combustion engines as range extenders, allowing them to recharge batteries mid-flight or supplement propulsion during takeoff and climb.

8.2.8 Advantages of Range Extenders and Turboelectric Concepts

- Extend operational range beyond battery limitations
- Enable regional and short-haul electric flights
- Reduce carbon emissions and fuel consumption
- Operate efficiently at constant turbine speeds
- Support distributed propulsion designs
- Lower noise and vibration levels
- Provide redundancy and improved system reliability
- Allow use of alternative fuels like hydrogen and SAFs
- Facilitate gradual transition to full electrification
- Expand payload and endurance capabilities

8.3 FUEL CELL-POWERED AVIATION

8.3.1 Core Technological Principles and System Architecture

The operational foundation of fuel cell aviation rests on sophisticated electrochemical processes and integrated system designs that must function reliably under demanding flight conditions. Proton Exchange Membrane Fuel Cells represent the most prevalent technology for aviation applications, valued for their high efficiency, rapid start-up capabilities, and suitability for the dynamic power demands of aircraft operations.

These fuel cells operate through an electrochemical reaction where hydrogen and oxygen combine to produce electricity, with water and heat as the only outputs. The complete fuel cell system extends beyond the core stack to include an essential Balance of Plant comprising supporting components such as air compressors, humidification systems, thermal management units, and power electronics. Optimizing this Balance of Plant is critical for achieving the power-to-weight ratios necessary for viable aircraft performance, as these auxiliary components can significantly impact overall system efficiency and reliability.

Hydrogen storage and handling constitute equally crucial technological challenges for aviation implementation. For aircraft applications, hydrogen is typically stored in liquid form at cryogenic temperatures to maximize energy density while minimizing storage volume. While liquid hydrogen offers substantial weight advantages compared to conventional fuels, it requires approximately four times the volume for equivalent energy content, presenting significant aircraft design challenges regarding tank integration and aerodynamics. Advanced storage systems incorporate multiple safety features including double-walled insulated tanks, comprehensive leak detection sensors, and active ventilation systems to manage hydrogen's specific safety considerations. Ongoing research focuses on developing advanced composite materials for cryogenic tanks and improving the gravimetric efficiency of storage systems to increase the proportion of usable hydrogen relative to total system weight.

8.3.2 Current Market Landscape and Growth Projections

The aircraft fuel cell market is experiencing accelerated development driven by environmental regulations, technological advancements, and growing industry commitment to decarbonization. Market analysts project substantial growth trajectories as the technology matures and demonstrates commercial viability. The most optimistic forecasts anticipate expansion from several billion dollars to over twenty billion dollars within a decade, reflecting compound annual growth rates exceeding twenty percent. This growth pattern underscores the aviation industry's increasing confidence in fuel cell technology as a primary solution for regional and commuter aircraft categories. The market segmentation reveals distinct trends across applications and platforms that highlight the technology's evolving role in aviation. In application terms, the market is currently led by Auxiliary Power Units that provide electrical power for aircraft systems during ground operations, reducing the need to run main engines while parked. However, the most dynamic growth is occurring in main propulsion systems, indicating the industry's transition from using fuel cells for supplementary power to employing them as primary propulsion sources. From a platform perspective, Unmanned Aerial Vehicles currently represent a significant market segment due to lower certification barriers and early adoption in defense and commercial applications. The most rapid expansion is projected for the Advanced Air Mobility segment, particularly electric vertical takeoff and landing aircraft designed for urban air mobility services. Geographically, North America maintains leadership in technology development supported by regulatory frameworks and established aerospace manufacturing capabilities, while the Asia-Pacific region demonstrates the fastest growth rate fueled by national hydrogen strategies and expanding aviation markets.

8.3.3 Industry Participants and Development Initiatives

The development of fuel cell aviation involves a diverse ecosystem of established aerospace manufacturers, specialized technology companies, and innovative startups collaborating to advance the technology toward commercial implementation.

- **ZeroAvia:** A prominent developer focused on hydrogen-electric powertrains for regional aircraft, targeting certification for progressively larger power systems and securing conditional orders from multiple airlines.
- **Intelligent Energy:** A leading fuel cell technology provider participating in major aerospace research initiatives and developing specialized fuel cell modules for aviation applications.
- **H_2FLY:** A German company that achieved the world's first piloted flight of a liquid hydrogen-powered electric aircraft, demonstrating extended endurance capabilities.
- **Cranfield Aerospace Solutions:** Pursuing aircraft conversion strategies to retrofit existing regional aircraft with hydrogen fuel cell propulsion systems.
- **PowerCell Group:** Engaged in European research consortia focused on scaling fuel cell power capacity and optimizing system components for aviation requirements.
- **H_3 Dynamics:** Advancing high-power aviation fuel cell stacks with recent certifications demonstrating substantial increases in power output.
- **Embraer:** Developing new aircraft concepts specifically designed for hydrogen fuel cell propulsion as part of its sustainable aviation technology roadmap.

- **Airbus:** Collaborating on megawatt-class fuel cell development programs targeting larger aircraft categories through specialized joint ventures.
- **Fokker NextGen:** Working to convert regional jet aircraft to liquid hydrogen propulsion in hybrid configurations with subsequent clean-sheet aircraft designs.
- **Academic Research Institutions:** Universities and research organizations are exploring novel fuel cell concepts including alternative chemistries that could offer improved energy density and safety characteristics.

8.3.4 Technical Challenges and Implementation Barriers

Despite significant progress, the widespread adoption of fuel cell aviation faces substantial technical, infrastructural, and economic challenges that must be systematically addressed. The high cost of fuel cell systems represents a primary economic barrier, with aviation-grade stacks currently requiring precious metal catalysts and precision manufacturing that elevate costs substantially compared to conventional propulsion systems. The supporting balance of plant components, including cryogenic storage systems and specialized compressors, also lack established supply chains, limiting opportunities for cost reduction through economies of scale. These cost factors create extended payback periods that can discourage airline investment despite potential operational savings.

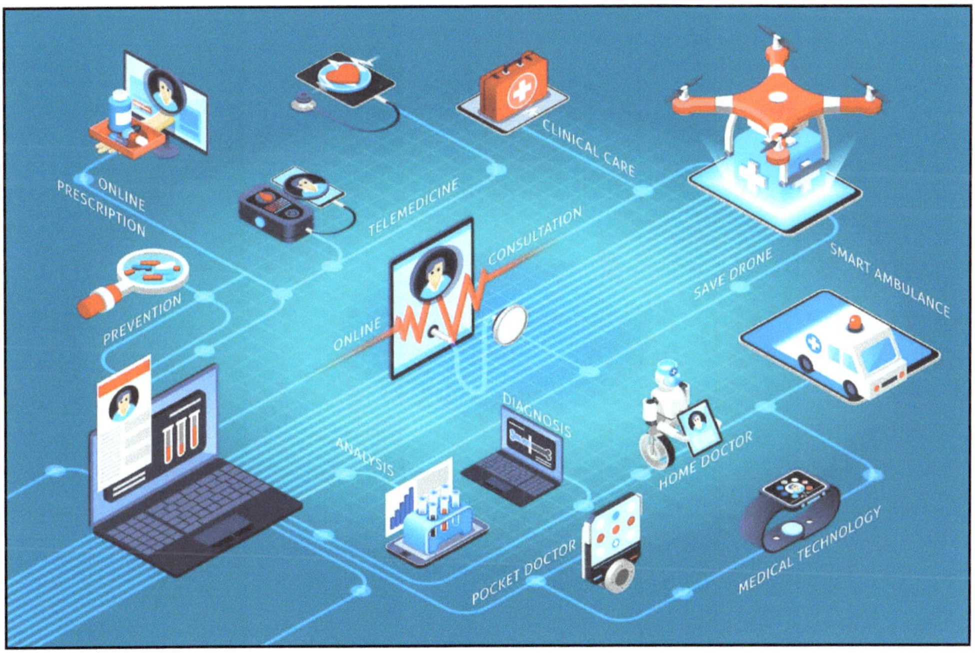

Hurdles in Implementation

The limited availability of liquid hydrogen infrastructure at airports presents another critical implementation challenge. Very few airports worldwide currently possess the capability to handle and dispense liquid hydrogen, which requires specialized cryogenic storage, transportation, and refueling equipment.

Developing this infrastructure demands substantial capital investment that can be difficult to justify without a critical mass of hydrogen-powered aircraft, creating a classic coordination problem between aircraft operators and infrastructure providers. Simultaneously, significant technical challenges remain in scaling the technology for larger aircraft categories. While fuel cells demonstrate excellent characteristics for small to midsize aircraft, scaling to megawatt-class power levels required for regional jets involves overcoming hurdles in thermal management, system integration, and maintaining performance at high altitudes where air density decreases. The regulatory and certification framework continues to evolve as aviation authorities develop comprehensive standards for these new propulsion technologies, a necessary but time-consuming process that adds complexity to development timelines.

8.3.5 Environmental Impact and Sustainability Considerations

The environmental implications of fuel cell aviation represent a central driver for technology development and a critical consideration in assessing its potential role in sustainable air transportation. The most significant environmental advantage of hydrogen fuel cells is the elimination of carbon dioxide emissions during flight operations, addressing aviation's contribution to climate change. When produced using renewable energy sources, hydrogen enables truly zero-emission flight operations throughout the entire energy lifecycle. Beyond carbon dioxide, fuel cell systems produce minimal air pollutants compared to conventional aircraft engines, potentially reducing impacts on local air quality near airports and along flight paths. The technology also offers substantial noise reduction benefits compared to traditional propulsion systems, potentially enabling more flexible operating patterns and reduced community noise impacts.

The sustainability of fuel cell aviation depends critically on the production methods for hydrogen fuel. Currently, most hydrogen production relies on fossil fuels through processes that generate significant carbon emissions, which would substantially undermine the environmental benefits of fuel cell aircraft. Transitioning to green hydrogen production using renewable electricity is essential for realizing the full climate potential of the technology. The water vapor emissions from hydrogen combustion at high altitudes present additional environmental considerations that require further scientific understanding regarding potential climate impacts. Lifecycle assessment methodologies must be applied to evaluate the comprehensive environmental footprint of fuel cell aircraft, including manufacturing processes, infrastructure requirements, and end-of-life considerations. As the technology develops, establishing sustainable practices for fuel cell production, operation, and eventual decommissioning will be essential to ensure that environmental benefits are realized across the complete technology lifecycle.

8.4 HYDROGEN-ELECTRIC INTEGRATION

Hydrogen-electric integration is a pragmatic bridge between today's battery-limited range and tomorrow's long-haul decarbonization. It packages hydrogen's outstanding gravimetric energy into an electric architecture that is quiet, responsive, and compatible with distributed propulsion. The technology pieces—efficient PEM stacks, robust BoP, crashworthy tanks, SiC power electronics, high-specific-power motors, and smart thermal systems—are all advancing, and early regional demonstrators point toward compelling payload–range and operating economics as hydrogen supply scales and green pathways proliferate. Success will come from system craftsmanship rather than any single component breakthrough: matching hydrogen form to mission, integrating thermal and air management tightly with power electronics, designing for safety and maintainability from the start, and building ground ecosystems that deliver clean hydrogen reliably and affordably.

If aviation is to decarbonize on credible timelines, a portfolio approach is essential. Batteries will dominate the shortest routes; hydrogen-electric will extend the electric envelope into regional missions; sustainable fuels will cover longer sectors until other options mature. The common thread is electrification know-how—high-voltage safety, thermal integration, predictive energy management—that hydrogen-electric integration both demands and accelerates. With disciplined engineering and coordinated infrastructure, hydrogen and electrons can fly farther together than either can alone.

8.4.1 Architectures: How Hydrogen Meets Electrons

The most common path is fuel-cell electric propulsion, where hydrogen reacts with oxygen electrochemically to produce DC power and water, typically via proton exchange membrane (PEM) stacks operating near 60–80°C. The fuel cell provides cruise energy at high efficiency, while a battery buffers transients for takeoff, climb, go-arounds, and regenerative absorption in descents for some concepts. Power flows over a high-voltage DC bus to inverters that drive distributed motors. A second family is hydrogen hybridization: hydrogen-fueled internal combustion engines or turbogenerators produce electric power in a series hybrid configuration, using the battery for peaks. This approach leverages familiar rotating machinery and can be less sensitive to stack air and humidity management, though it reintroduces hot sections and NOx unless mitigated. A third, longer-horizon pathway is solid oxide or high-temperature PEM stacks for improved tolerance to impurities and higher-grade waste heat, but these bring warm-up time and materials challenges that favor larger aircraft.

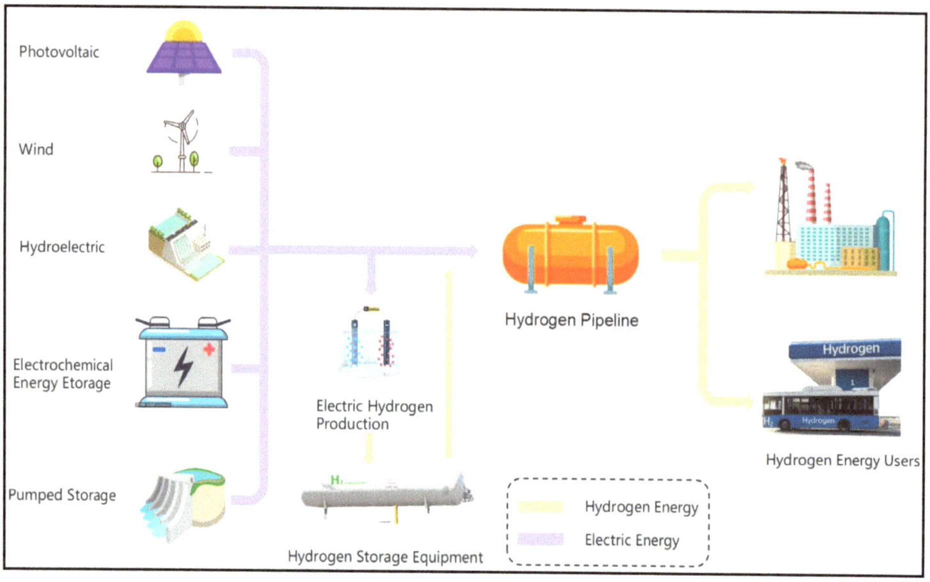

Intersection of Hydrogen and Electric Power

In all cases, batteries remain integral. They cover fast-changing loads that would incur fuel-cell compressor lag or stack stress, enable silent taxi, and provide emergency reserves. Pack sizing is an economic and safety balance: large enough for performance and contingencies, small enough to avoid unnecessary mass and cost.

8.4.2 Hydrogen Storage Choices: Compressed Gas vs. Liquid

Hydrogen's gravimetric specific energy is exceptional, but its volumetric energy density is low. Compressed gaseous hydrogen (CGH2) at 350–700 bar uses carbon composite cylinders with high safety factors. It avoids cryogenics and simplifies ground handling but demands significant fuselage volume and aerodynamic integration to manage drag and center-of-gravity shifts. Liquid hydrogen (LH2) increases volumetric density by roughly an order of magnitude versus low-pressure gas by cooling to about -253°C.

LH2 tanks are double-walled, vacuum-insulated dewars with minimal boil-off but stringent thermal and material requirements. The mass fraction of tanks to stored hydrogen improves with size, which favors regional and larger aircraft for LH2, while CGH2 can suit smaller platforms where simpler operations trump range. Infrastructure and lifecycle energy also matter. Liquefying hydrogen consumes significant energy—often on the order of a third of hydrogen's lower heating value—while compressing to 700 bar is materially lower but still nontrivial.

Airports can choose pathways based on supply chains: delivered LH2 by tanker, on-site electrolysis and compression, or hybrid hubs that accept multiple forms. These choices reverberate through turnaround times, safety zones, and economics.

8.4.3 Balance of Plant and Power Electronics

Fuel cells require a "balance of plant" (BoP) that is as critical as the stack. Air management uses electrically driven turbo-compressors to deliver pressurized, filtered air to the stack; their efficiency and responsiveness directly affect net power and dynamics. Humidification maintains membrane conductivity without flooding, and purge cycles prevent nitrogen buildup and manage water balance. Hydrogen recirculation or ejectors improve fuel utilization, while pressure regulators and isolation valves ensure stable, fault-tolerant flow. Sensors for temperature, pressure, humidity, and hydrogen concentration feed control loops governed by safety and performance priorities. On the electrical side, a DC/DC converter conditions fuel-cell output to the high-voltage bus, decoupling stack dynamics from load steps. Batteries connect via bidirectional converters, allowing both assist and recharge. Inverters drive motors at variable frequency, with silicon carbide devices increasingly preferred for high efficiency and compact cooling. Protection schemes include fast-acting contactors, arc-fault and ground-fault detection, and segmented buses to isolate faults without collapsing power to critical loads. The supervisory controller orchestrates the split between fuel cell and battery to keep the stack in efficient operating zones while ensuring transient responsiveness.

8.4.4 Thermal Integration: Heat is the Hidden Constraint

Every watt of electrical output from a PEM fuel cell is accompanied by roughly comparable waste heat at modest temperature, which must be rejected to the airstream or absorbed by thermal buffers. At low altitude and moderate speed, air-cooled radiators and liquid loops can suffice; at high altitude or for low-speed eVTOL operations, heat rejection becomes harder and can dominate drag and mass. Larger heat exchangers, variable-speed fans, and even phase-change materials to ride through hover peaks are part of the toolkit. The battery, inverters, and motors add their own thermal loads, often at higher allowable temperatures, creating an opportunity for cascaded loops where a high-temperature loop preheats a low-temperature loop or vice versa via heat pumps. Thoughtful thermal integration reduces radiator size, improves overall efficiency, and, crucially, protects stack life by avoiding hot spots, membrane dehydration, or freeze/thaw damage.

8.4.5 Safety, Sensing, and Certification Considerations

Hydrogen brings specific hazards—wide flammability range, low ignition energy, and embrittlement concerns—that are managed through design. Tanks are located within protected fuselage zones or external pods engineered for crash loads and penetration resistance. Vent lines route any intentional or accidental releases to controlled overboard locations, leveraging hydrogen's buoyancy for rapid dispersion. Comprehensive leak detection at joints and enclosures, continuous ventilation of compartments, and graded isolation with automatic valve closures are standard practices. Materials are selected for compatibility at cryogenic temperatures for LH2 and to resist hydrogen permeation.

Certification frameworks for hydrogen-electric systems are maturing. Authorities expect applicants to apply standard system safety processes, environmental testing, and electromagnetic compatibility verification, supplemented by hydrogen-specific analyses of leak scenarios, ventilation, ignition sources, and crashworthiness. Fuel-cell stacks and BoP undergo endurance testing across altitude, temperature, and humidity ranges, and emergency procedures include rapid shutdown and safe venting. Battery integration remains under familiar rules but must show safe interaction with hydrogen systems, especially in shared compartments and under fault conditions.

8.4.6 Operations and Airport Infrastructure

Turnarounds define viability. CGH2 refueling uses high-pressure connectors and staged banks to limit temperature rise during fast fills; LH2 requires cryogenic couplings, purging, and disciplined procedures to avoid air condensation and embrittlement. Regardless of form, clear safety zones, trained crews, and gas detection around fueling points are prerequisites. Airports that already host industrial gas operations have a head start, but widespread adoption will demand co-designed layouts that separate hydrogen from passenger flows while minimizing taxi and ground time. On-site electrolysis marries well with airport microgrids, converting off-peak or renewable electricity into hydrogen and smoothing demand on the grid. Liquefaction on site remains capital intensive but could make sense for large hubs with heavy hydrogen demand.

Coordination with aircraft energy management systems improves efficiency. Preheating or prehumidifying stacks reduces startup penalties, while predictive fueling aligns delivered mass with planned missions and reserves, minimizing boil-off or residual pressure. Digital fueling records, quality assurance for fuel purity, and standardized communications between ground systems and aircraft close the loop on safety and performance.

8.4.7 Economics and Sustainability

Hydrogen-electric integration competes on cost per available seat-kilometer, reliability, and environmental credentials. The big variables are hydrogen price at the airport fence, stack and BoP lifetime, battery replacement cadence, and utilization. Green hydrogen from renewable-powered electrolysis carries a premium today but is trending downward as electrolyzer scale grows and renewable costs fall. Because fuel cells operate efficiently at steady loads, hybridization that assigns peaks to batteries can prolong stack life and reduce fuel costs. Maintenance for electric drivetrains remains low relative to combustion engines, while stack overhaul intervals and membrane durability are improving through better water management and contaminants control.

Lifecycle emissions depend on hydrogen's production pathway. Green hydrogen can deliver near-zero operational CO_2; transitional "blue" hydrogen reduces lifecycle impacts if coupled with high-capture, low-leakage carbon management. Material circularity is advancing via fuel-cell catalyst recycling and composite tank reuse or requalification. As with batteries, durable design, second-life opportunities, and recycling infrastructure will shape the true sustainability of hydrogen aviation.

8.4.8 Cases and Mission Fit

Hydrogen-electric excels where energy demand exceeds what near-term batteries can support without onerous mass penalties, yet where the speed and altitude simplify heat rejection and air management. Regional fixed-wing aircraft carrying 9–40 passengers over 300–800 km are prime candidates, with LH2 unlocking the best payload-range balance. Cargo feeders benefit similarly, trading cabin integration complexity for volume flexibility. In eVTOL, hydrogen can extend range and cycle life, but thermal management during hover is challenging; lift-plus-cruise architectures that minimize hover time are better suited. High-altitude, long-endurance UAVs already demonstrate solar-hydrogen hybrids: solar arrays power daytime flight and electrolyze hydrogen on the ground or store it from prior production, with fuel cells carrying the night—an archetype for endurance-driven missions.

8.4.9 Integration Challenges Worth Solving

Three integration themes dominate. First, volumetric packaging and aerodynamics: tanks must be placed to preserve lift distribution, minimize drag, and maintain center-of-gravity across fuel burn. Blended wing-body and high-volume fuselage concepts help, as do external pods for retrofits. Second, dynamic performance: ensuring smooth, reliable power through rapid transitions requires fast DC/DC control, predictive dispatch between stack and battery, and robust air management to prevent compressor stall or membrane stress. Third, heat: sizing radiators for worst-case hover or high-altitude climb without crippling cruise efficiency is a delicate balance that rewards phase-change buffers, variable geometry, and smart scheduling of thermal loads.

8.4.10 Design and Operations Levers for Hydrogen-Electric Integration

- Select the hydrogen form to match mission and scale: CGH2 for simplicity and smaller airframes, LH2 for regional range and better system-level density as aircraft size grows.
- Hybridize wisely: use the fuel cell as a steady cruise engine and size batteries for take-off/climb peaks, go-arounds, and contingencies to extend stack life and raise net efficiency.
- Engineer air management for altitude: high-efficiency, fast-response compressors with adequate surge margin and humidity control maintain stack performance in thin, cold air.
- Treat heat as a first-class constraint: multi-loop cooling with shared buffers, smart radiator placement, and phase-change storage rides through hover/climb without excessive drag.
- Package tanks with aerodynamics in mind: integrate conformal tanks or external pods that preserve lift distribution, minimize interference drag, and keep CG within limits across burn.
- Harden safety layers: hydrogen detection, forced ventilation, graded isolation valves, and controlled vent masts—backed by fault trees and flight-tested emergency procedures.
- Standardize high-voltage architecture: segmented DC buses, fast fault detection, and SiC-based power electronics reduce losses and maintain stability under rapid load changes.
- Optimize fueling and quality control: filtered, dry hydrogen meeting aviation-grade purity, standardized connectors, digital custody chains, and procedures that minimize turnaround scatter.
- Use predictive energy management: couple weather, altitude, and mission profiles to dispatch fuel cell vs. battery power, protect reserves, and precondition stacks and packs on the ramp.
- Build airport microgrids around hydrogen: co-locate electrolysis, storage, and DC fast charging to arbitrate between electrons and molecules, cut carbon intensity, and add resilience.

8.5 COMPARATIVE PERFORMANCE AND MARKET APPLICATIONS

Comparative performance and market applications are two sides of the same coin. Series hybrids convert their architectural decoupling into standout results in stopstart urban work, smooth torque delivery, and simplified drivelines that suit buses, delivery fleets, rail, and marine. Parallel hybrids turn mechanical continuity into superior highway efficiency, familiar drivability, and cost-effective integration in passenger cars and mixed-use fleets. Plugin capability blurs boundaries by injecting substantial electric miles into daily operation, with outcomes governed by charging access and grid cleanliness.

Battery-electric vehicles set a high bar in urban and depot-centric duty and will continue to expand as infrastructure and batteries improve, while fuel cells target heavy-duty corridors as hydrogen ecosystems mature. The market will remain plural because duty cycles differ, but the common denominator is system optimization: matching hardware to mission, orchestrating energy with intelligent software, and building supply chains and recycling streams that make environmental and economic sense over the full life of the vehicle. In that alignment, comparative performance stops being a contest of architectures and becomes a design discipline tuned to the work vehicles actually do.

8.5.1 Comparative Performance Actually Means

Performance in context starts with the energy equation. Energy consumed per mile or per tonne-kilometer is the anchor for operating cost and carbon intensity, but it is only the opening act. Emissions profiles add local pollutants such as nitrogen oxides and particulates that matter most in urban corridors and around depots. Drivability covers torque response, launch feel, shift quality, and brake blending, which shape user satisfaction and safety. Noise, vibration, and harshness influence brand identity and passenger comfort, particularly in premium segments and public transit. Uptime and maintenance determine whether promised savings survive the realities of wear and tear, while thermal resilience under hot summers, alpine grades, or towing sets the limits of repeatable performance. Total cost of ownership stitches these threads together by including purchase price, energy, service, insurance, residual value, and incentives, which often make the difference between a compelling pilot and a scalable fleet deployment.

8.5.2 Duty Cycle Sensitivity across City, Suburb, and Highway

The purest separator among propulsion choices is duty cycle. In dense urban traffic with frequent stops, architectures that harvest braking and minimize engine transients excel. Series hybrids that decouple the engine from the wheels keep the combustion unit in efficient islands or off entirely, while traction motors handle acceleration and regeneration. The cumulative gains add up quickly in buses, refuse trucks, and last-mile vans that rarely see prolonged high-speed cruising. In mixed suburban conditions, where arterial roads alternate with short highway bursts, parallel hybrids leverage a direct mechanical path at cruise while using electric torque for launches and brief EV operation. On sustained highway routes, the advantage tilts toward mechanically efficient layouts that avoid repeated energy conversions, which is why parallel or power-split systems typically outperform series configurations at steady speeds. Battery-electric vehicles can be exceptional in urban settings where range is predictable and regeneration is abundant, but must carry more storage or rely on faster charging to match the same confidence at extended highway speeds.

8.5.3 PlugIn Threshold and Grid Interactions

Adding plug-in capability reshapes comparative outcomes by shifting a large share of miles to electricity. For commuters with reliable home or depot charging, even modest battery capacities allow most daily trips to be electric, with the combustion engine acting as a long-range safety net. Plugin parallel hybrids pair this electric everyday experience with efficient mechanical drive on longer journeys, while plugin series hybrids deliver a consistently electric feel and rely on a generator for sustained power once the battery depletes. Grid carbon intensity and charging behavior govern lifecycle emissions and cost. Clean electricity amplifies the advantage of plug-ins and battery-electric vehicles, while coal-heavy grids can blunt benefits unless charging is timed to renewable availability or sourced via green tariffs. Software increasingly orchestrates charging to off-peak hours and aligns it with fleet schedules, turning energy procurement into a performance lever grounded in both economics and sustainability.

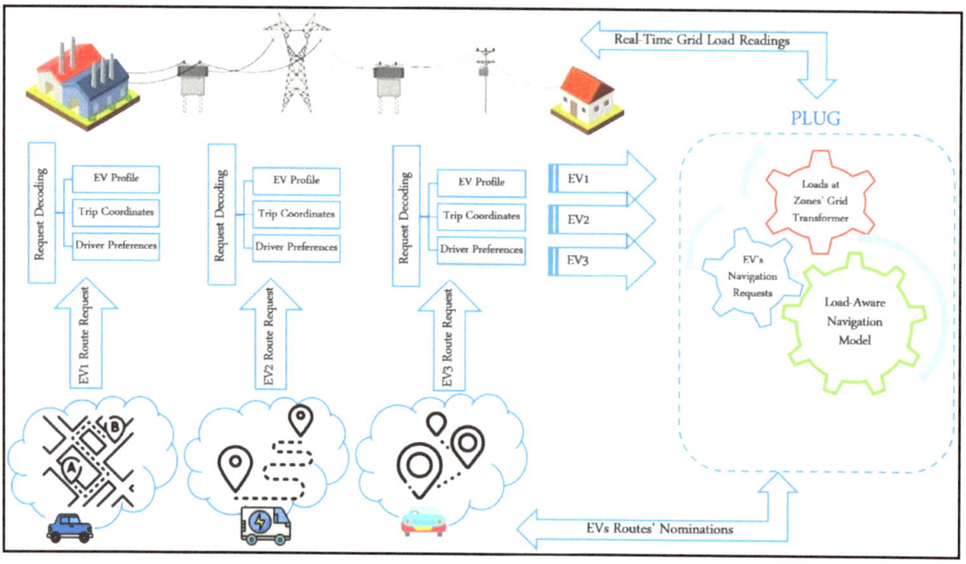

Connection Limits and Power Grid Interactions

8.5.4 Passenger Vehicles: Sedans, Crossovers, and Premium

In the mainstream car market, breadth of ability often beats specialization. Consumers expect strong fuel economy without sacrificing acceleration, quietness, or long-distance comfort. Parallel hybrids integrated into dedicated transmissions provide a pragmatic balance by enabling engine downsizing, smoothing shifts with electric torque fill, and preserving low-loss mechanical drive at highway speeds. They slot into existing platforms, benefit from mature supply chains, and deliver predictable behavior across climates and topographies. Plugin variants extend the appeal for buyers with access to charging, offering meaningful electric range for daily errands and efficient blended operation for road trips.

Series-style passenger offerings find their footing in urban-focused models that emphasize an electric driving feel, simplified driveline components, and low NVH, where high-speed efficiency penalties are less consequential. Battery-electric cars increasingly set the benchmark for refinement and performance, and where public charging is dense and reliable, they can displace hybrids entirely, but hybrids remain compelling for single-vehicle households facing occasional long journeys or limited charging access.

8.5.5 Commercial Fleets: Buses, Delivery, and LongHaul

Fleet operators judge technology primarily by total cost and uptime. City buses and refuse haulers are often the first to adopt electrification because their routes are short, predictable, and regeneration-rich, and because depots simplify charging logistics. Series hybrids have been a stepping stone in these segments, offering smoother torque control, reduced local emissions, and easier maintenance by eliminating multi-speed gearboxes. Battery-electric buses now increasingly reach cost parity over their lifetimes, buoyed by incentives and lower energy and maintenance costs, and they enjoy strong public support due to quiet operation and zero tailpipe emissions. Urban delivery fleets straddle mixed duty cycles; here, series hybrids or full battery-electric vans excel in dense zones, while parallel hybrids perform well on suburban and exurban stretches with higher average speeds. Longhaul trucking highlights the limits of hybridization because sustained highway cruising leaves little kinetic energy to recover, tilting the field toward high-efficiency combustion with renewable fuels today and toward hydrogen or battery-electric solutions on defined corridors as megawatt charging and hydrogen networks mature.

8.5.6 OffHighway, Rail, and Marine: Specialized Duty

Specialized sectors have long exploited electric drive for its controllability and modularity. Diesel-electric locomotives and many marine vessels effectively operate as series hybrids, running engines at steady, efficient points while traction motors provide high torque and precise control without complex gearboxes. The addition of batteries turns these into hybrids capable of peak shaving, silent ingress and egress in environmentally sensitive zones, and reduced engine idling. Construction and mining equipment increasingly integrate hybrid functions to recover energy from lowering loads and braking or to smooth power spikes from implements, with the choice between series and parallel arrangements determined by the balance of traction versus auxiliary demands. In these markets, reliability, ease of service in remote locations, and parts commonality often outweigh marginal efficiency differences, which is why proven architectures with robust thermal and dust management win contracts even if laboratory fuel savings are similar across options.

8.5.7 Regional Dynamics and the Role of Policy

Geography and policy sculpt market adoption patterns. European fleet CO_2 targets, urban low-emission zones, and high fuel prices have made hybrids and plugins standard fare in many segments, while rapid charging buildout and strong consumer incentives accelerate battery-electric uptake. China's dual-credit system, localized supply chains, and aggressive infrastructure investment have propelled electrification broadly, though hybrids remain important compliance tools and bridge technologies in areas where charging density lags.

Japan's hybrid leadership reflects dense urban driving combined with periurban travel, favoring power-split and parallel architectures for their balanced efficiency and familiar refueling experience. North America's long-distance travel culture and climate diversity sustain demand for solutions that combine highway efficiency with cold-weather resilience, positioning parallel hybrids and plugins as pragmatic choices even as the fast-charging network expands. Stable, long-horizon regulations and procurement policies are decisive because they unlock manufacturing scale and lower risk for fleets, ultimately turning technical merit into commercial reality.

8.5.8 Economics, Reliability, and Supply Chain Resilience

Comparative economics hinge on component costs, manufacturing complexity, service needs, and residual values. Parallel hybrids often capitalize on existing engines and transmissions augmented by electric machines and clutches, achieving scale quickly with incremental platform changes. Their batteries can be smaller in non-plugin forms, containing costs and preserving cargo space. Series hybrids simplify mechanical drivetrains but require larger traction motors, generators, and robust power electronics, shifting value into copper, magnets, and thermal management.

Trends in lithium pricing, rare-earth magnet availability, and the adoption of silicon carbide semiconductors influence the balance. Reliability is multidimensional: parallel hybrids demand precise control to blend torque and manage clutches, while series systems reduce moving parts in the driveline but concentrate stress in electrical components that must withstand high cycling and heat. Warranty strategy, dealer training, and parts logistics matter as much as headline technology, especially for fleets that cannot tolerate downtime. Recycling pathways for batteries and magnets are emerging competitive differentiators as regulations begin to reward recycled content and penalize unmanaged waste.

8.5.9 Competition and Complementarity with BEVs and Fuel Cells

Battery-electric platforms are the most disruptive comparator because they remove the engine entirely and deliver unmatched smoothness and often lower operating costs, particularly in urban and depot-based use. As batteries become cheaper and denser, and as charging infrastructure improves in reliability and speed, BEVs will continue to encroach on territory once held by hybrids. That said, hybrids remain powerful in scenarios where range flexibility, rapid refueling, extreme climates, or towing are common, and where customers prefer minimal change to operating patterns. Fuel cell electric vehicles, essentially series electric drives with hydrogen as the energy carrier, offer high specific energy and fast refueling that could fit heavy-duty, long-range use if hydrogen production, distribution, and stack durability targets are met. In the near term, hybrids and BEVs are more complements than competitors in portfolio strategy, allowing manufacturers to address diverse customer needs and regulatory requirements while hedging against infrastructure uncertainty.

8.5.10 Technology Trajectories and Strategic Positioning

The direction of travel favors convergence and software-defined performance. Power-split transmissions blend series and parallel benefits and keep expanding their efficient operating envelope, while wide-bandgap semiconductors reduce conversion losses and shrink thermal systems, narrowing penalties for electric energy paths at higher speeds. Battery advances, including higher nickel chemistries, lithium iron phosphate in cost-sensitive segments, and emerging solid-state designs, extend electric ranges for plugins without punishing mass, shifting usage toward electric miles by default. Predictive energy management uses maps, traffic, weather, and topography to schedule engine operation, precondition batteries for efficiency and longevity, and optimize charging to low-carbon windows, turning data into real-world savings. Strategically, the strongest market positions will belong to platforms that scale across multiple propulsion variants, share components broadly, and are supported by transparent lifecycle assessments that align with procurement rules and consumer expectations.

CHAPTER 9

Urban Air Mobility (UAM) and Advanced Air Mobility (AAM)

9.1 CONCEPT AND VISION OF UAM

The concept and vision of Urban Air Mobility represents one of the most significant transportation innovations of our time. By combining advanced technology with visionary urban planning, UAM promises to create more efficient, sustainable, and livable cities while generating substantial economic and environmental benefits. As this ecosystem continues to develop and mature, UAM will increasingly shape how we understand and approach urban transportation in the decades to come.

9.1.1 Core Concept of Urban Air Mobility

At its foundation, Urban Air Mobility operates on a relatively straightforward principle: utilizing airspace above cities as a viable transportation corridor. Rather than relying exclusively on ground-based infrastructure such as roads, highways, and public transit systems, UAM proposes that aircraft can efficiently move passengers and cargo through the three-dimensional space above urban environments.

This concept leverages recent technological advances in electric propulsion systems, autonomous navigation, battery technology, and air traffic management systems that have made such operations increasingly feasible. The core appeal of UAM lies in its potential to bypass surface-level congestion, dramatically reducing travel times for both passengers and time-sensitive cargo, while simultaneously decreasing the burden on existing ground transportation infrastructure.

9.1.2 Current Technological Foundations

The technological backbone supporting UAM development includes several key innovations that have matured over recent years. Electric vertical takeoff and landing vehicles, often referred to as eVTOLs, represent the primary aircraft type being developed for UAM operations. These vehicles can operate from rooftops, parking lots, and specially designated vertiports, eliminating the need for extensive runway infrastructure that traditional aircraft require. Advanced battery technologies, particularly improvements in lithium-ion and next-generation solid-state batteries, have made extended flight durations increasingly possible.

Additionally, autonomous flight systems continue to advance rapidly, incorporating sophisticated sensors, artificial intelligence, and machine learning algorithms that enable safe and efficient operation with minimal or reduced human intervention. These technological components work together to create a viable foundation for UAM operations in complex urban environments.

9.1.3 Vision for Future Urban Transportation

The overarching vision for UAM encompasses a comprehensive reimagining of urban transportation networks. This vision projects a future where millions of people can move seamlessly through cities using a combination of ground-based and air-based transportation modes, with decisions about which mode to utilize made based on efficiency, cost, and environmental impact. Early conceptualizations suggest that UAM could handle express routes between distant points in cities, connecting to traditional public transportation networks at designated vertiports located strategically throughout metropolitan areas. The vision emphasizes seamless integration with existing mobility services, potential connections to airports for longer-distance travel, and optimization of the entire transportation ecosystem through advanced data analytics and real-time traffic management systems.

Prospects for Next-Generation Urban Transit

9.1.4 Sustainability and Environmental Benefits

One of the most compelling aspects of the UAM vision concerns its potential environmental advantages. Electric-powered aircraft generate zero direct emissions during flight, offering substantial improvements over conventional automotive transportation that relies on combustion engines. When combined with renewable energy sources powering the charging infrastructure, UAM could dramatically reduce the carbon footprint associated with urban transportation. Additionally, the efficiency gains achieved through reduced congestion and optimized routing can contribute to lower overall energy consumption per passenger-mile traveled. The vision for UAM incorporates these environmental benefits as central to its value proposition, particularly as cities worldwide commit to achieving net-zero emissions targets and addressing climate change concerns. This alignment with global sustainability objectives makes UAM an attractive solution for forward-thinking urban planners and policymakers.

9.1.5 Economic and Operational Efficiency

The economic vision of UAM centers on the creation of significant value through time savings, operational efficiency, and the emergence of entirely new industries and employment opportunities. By reducing travel times for both passengers and goods, particularly in congested urban areas, UAM could yield substantial economic benefits through increased productivity and reduced logistics costs. The concept envisions a mature market where competition among multiple operators drives innovation and reduces costs, eventually making UAM services accessible to broader segments of the population. Furthermore, the operational efficiency inherent in autonomous flight systems and optimized air traffic management promises lower per-unit operating costs compared to ground-based alternatives. The vision includes the development of supporting infrastructure, maintenance facilities, and specialized workforce training programs that would generate significant employment and economic activity.

9.1.6 Urban Planning and Infrastructure Integration

The UAM vision fundamentally challenges and reshapes how urban planners approach metropolitan development and transportation infrastructure. This emerging concept requires the development of a new class of urban infrastructure, including vertiports designed for passenger convenience and efficient aircraft turnaround. The vision includes strategic placement of these facilities to maximize accessibility and minimize noise and safety concerns for urban residents. Additionally, UAM's integration into broader urban planning strategies promises to influence land use patterns, potentially reducing the need for extensive ground-level transportation infrastructure and freeing valuable urban real estate for residential, commercial, or recreational purposes. This aspect of the UAM vision represents an opportunity to reimagine cities themselves, creating more livable, efficient, and sustainable urban environments that prioritize human well-being alongside transportation efficiency.

9.1.7 Safety, Regulation, and Governance

The vision for UAM incorporates robust safety frameworks and regulatory structures essential for public acceptance and operational viability. Advanced safety systems, including redundant propulsion capabilities, collision avoidance technologies, and automated emergency response protocols, represent critical components of the envisioned UAM ecosystem. The concept anticipates the development of comprehensive regulatory frameworks that address airspace management, pilot certification or autonomous system validation, noise limitations, and security considerations. The vision emphasizes collaborative governance involving aviation authorities, urban planners, environmental agencies, and private industry stakeholders working together to establish standards that ensure public safety while fostering innovation and sustainable growth.

9.2 DESIGN AND OPERATION OF EVTOL AIRCRAFT

The design and operation of eVTOL aircraft represent a monumental convergence of aerospace engineering, electrical systems, and digital infrastructure. Through innovative configurations and a relentless focus on lightweight and efficient design, engineers are solving the complex physics of electric vertical flight. In parallel, the development of vertiports and unmanned traffic management systems is building the necessary ecosystem for these aircraft to operate safely within cities. While significant hurdles in certification, battery technology, and public acceptance remain, the concerted effort from industry and regulators suggests that the era of urban air mobility is an imminent reality, poised to fundamentally reshape how people and goods move through the world.

9.2.1 Fundamental Design Principles and Aircraft Configurations

The core design philosophy of an eVTOL centers on achieving safe, quiet, and efficient flight within the demanding environment of a city. This mission is fundamentally enabled by electric propulsion, which relies on batteries, electric motors, and sophisticated power electronics to generate thrust. This system offers profound advantages over traditional internal combustion engines, including dramatically lower noise signatures, zero operational emissions, and higher mechanical reliability due to fewer moving parts.

A relentless focus on minimizing weight is paramount to extending the crucial metrics of flight time and payload capacity. Consequently, manufacturers extensively utilize advanced lightweight materials such as carbon fiber composites, high-performance polymers, and specialized aluminum alloys to achieve an optimal strength-to-weight ratio. The entire aircraft ecosystem, from its airframe to its avionics, is engineered around the principle of optimizing size, weight, and power to maximize performance. The industry has converged on several distinct design configurations, each representing a different engineering solution to the primary challenge of transitioning from vertical lift to efficient forward flight. The first and simplest is the multirotor design, which resembles a large drone with multiple fixed rotors. This configuration is valued for its mechanical simplicity and inherent stability during hover but is less efficient for longer-range cruise flight due to the drag of its stationary propellers. A step forward in complexity is the lift-plus-cruise configuration.

This design employs dedicated, separate sets of rotors—one set for vertical lift and another for forward propulsion. This allows each rotor system to be optimized for its specific task, improving overall efficiency. For higher performance and longer ranges, more complex solutions have been developed. The tiltrotor and tiltwing configurations feature rotors or entire wings that physically pivot, transitioning the direction of thrust from vertical for takeoff and landing to horizontal for forward flight. This approach combines the vertical lift capability of a helicopter with the efficient, high-speed cruise of a fixed-wing aircraft. Finally, the vectored thrust design utilizes a combination of fixed wings and rotors whose thrust direction can be manipulated through complex control systems. This provides excellent maneuverability and efficient cruise, though it requires extremely sophisticated software to manage the transition and flight phases. The choice of configuration is a fundamental trade-off between mechanical complexity, cruise efficiency, noise profile, and certification pathway.

9.2.2 Operational Ecosystem and Infrastructure

The successful deployment of eVTOL aircraft extends far beyond the vehicles themselves, requiring the development of an entirely new operational ecosystem. The cornerstone of this ground-based infrastructure is the vertiport. These are dedicated facilities for takeoff, landing, charging, and passenger processing. A vertiport is more than a simple helipad; it must include designated Touchdown and Lift-Off areas, passenger waiting and security zones, rapid charging stations, and basic maintenance facilities, all designed to support a high frequency of operations in a dense urban setting. The strategic placement of these vertiports to create a network that connects city centers, airports, and suburban hubs is a critical logistical and urban planning challenge.

Integrating eVTOLs safely into the existing airspace, particularly around busy cities, is one of the most significant operational hurdles. The current Air Traffic Management system was not designed to handle hundreds of small, low-altitude aircraft operating in close proximity. The solution being developed is a new, largely digital Unmanned Aircraft System Traffic Management system. This system will leverage advanced automation, real-time data sharing between aircraft and ground control, and dynamic deconfliction algorithms to manage flight paths, ensure safe separation, and prevent collisions. This creates a concept of "aerial highways" that can be managed with high efficiency and safety.

Functional Ecosystem and Structural Infrastructure

Commercial perspective, the primary operational model is Urban Air Mobility, which envisions a network of on-demand air taxi and ridesharing services. The initial use cases are focused on high-value routes such as airport shuttles, where an eVTOL could reduce a ground commute of over an hour to a flight of just a few minutes. Other early applications include emergency medical services, cargo delivery, and regional connectivity between nearby cities. The economic model relies on achieving a cost per passenger mile that is competitive with premium ground transportation options like ride-sharing services, while offering an unparalleled time saving. As the technology matures and scales, the vision is for these services to become increasingly accessible to the general public.

9.2.3 Critical Challenges in Certification and Scaling

The path to routine eVTOL operations is fraught with formidable challenges, the most immediate of which is the rigorous certification process. Aviation regulatory bodies have established the world's most stringent safety standards, and eVTOLs must meet these same high benchmarks. Manufacturers must demonstrate through extensive analysis and testing that their aircraft meets or exceeds the safety levels of conventional commercial aircraft. This process is complicated by the novel nature of the technology, including distributed electric propulsion and complex flight control software. Regulators are creating new specific certification standards for this "powered-lift" category, requiring exhaustive validation of every system, especially those providing redundancy for critical functions like propulsion and flight controls.

As the industry moves from prototypes to mass production, it faces significant scaling challenges. Battery technology remains a key limitation. While lithium-ion batteries are the current standard, the industry requires continued progress in energy density to support longer ranges and the demanding cycle of frequent take-offs and landings with quick recharge times. Safety protocols for high-capacity batteries in an aviation context, including thermal runaway containment, are also a primary focus of engineering efforts. Furthermore, the supply chain for advanced materials and specialized components is not yet matured for the volume required for mass production, which could constrain growth. Finally, a potential shortage of qualified pilots and maintenance technicians could create an operational bottleneck. In the longer term, the industry envisions fully autonomous flight to alleviate the pilot shortage and reduce costs, but public and regulatory acceptance of pilotless passenger flight remains a significant societal hurdle.

9.3 INFRASTRUCTURE, AIRSPACE INTEGRATION, AND TRAFFIC CONTROL

Infrastructure, airspace integration, and traffic control form the interdependent pillars of modern aviation. As global skies grow busier and more technologically complex, the aviation ecosystem must evolve toward automation, sustainability, and inclusivity. Seamless integration of digital tools, robust infrastructure, and adaptive regulatory frameworks will be essential to balance innovation with safety.

In the coming decades, the success of aviation will depend not only on how fast aircraft can fly but also on how intelligently they share the skies. The collaborative evolution of infrastructure and airspace management will define the efficiency, safety, and environmental integrity of the global aviation landscape—ushering in a new era of connected, sustainable air mobility.

9.3.1 Foundations of Aviation Infrastructure

Aviation infrastructure comprises a network of physical and digital systems that support the safe operation of aircraft. This includes airports, runways, communication systems, fueling stations, maintenance hangars, and navigation aids. Modern infrastructure planning also extends to ground transport connections and smart logistics. The efficiency of air operations depends on the quality and capacity of this infrastructure.

With increasing demand for air travel and emerging technologies such as electric vertical take-off and landing (eVTOL) vehicles, airports and vertiports must evolve. Infrastructure development must integrate renewable energy systems, advanced charging stations, noise reduction mechanisms, and automated maintenance technologies. Smart infrastructure also involves digital connectivity—enabling real-time data exchange between air and ground operations.

9.3.2 Emergence of Advanced Air Mobility (AAM) Systems

aviation is moving toward Advanced Air Mobility (AAM), which combines manned and unmanned aerial systems into a unified operational environment. AAM introduces urban air taxis, cargo drones, and regional electric aircraft that require specialized infrastructure such as vertiports, drone corridors, and rapid recharging networks.

Integrating these systems demands reimagining city planning and transport infrastructure. Urban centers must allocate air corridors, install landing zones on rooftops, and ensure the coexistence of aerial vehicles with traditional air traffic. The success of AAM depends on how effectively infrastructure can be adapted to accommodate diverse vehicle types without compromising safety and efficiency.

9.3.3 Airspace Integration: Challenges and Opportunities

Airspace integration refers to the seamless coordination of multiple aircraft types—commercial, private, and unmanned—within shared flight environments. Historically, airspace management was tailored for conventional aircraft flying at higher altitudes. The inclusion of drones and low-altitude operations introduces new layers of complexity. Challenges in airspace integration include collision avoidance, airspace congestion, data interoperability, and regulatory consistency across regions. To overcome these, advanced surveillance systems, satellite navigation, and digital communication networks are essential. The integration process also demands cooperation between civil aviation authorities, defense organizations, and private industry players. Opportunities emerge in the form of real-time airspace monitoring, predictive analytics, and automated flight management systems. With artificial intelligence (AI) and machine learning, airspace integration can transition from reactive to proactive management—anticipating congestion, rerouting traffic, and optimizing energy consumption.

9.3.4 Traffic Control and Airspace Management Systems

Air traffic control (ATC) serves as the backbone of aviation safety. It ensures that every aircraft in the sky maintains safe separation from others, follows assigned routes, and responds efficiently to weather or emergency conditions. Modern ATC systems are transforming from voice-based operations to digital, data-driven platforms. Automation, satellite-based tracking, and network-centric operations allow controllers to monitor and manage higher volumes of aircraft with improved accuracy. The introduction of the Automatic Dependent Surveillance-Broadcast (ADS-B) system, for instance, has revolutionized real-time aircraft tracking, reducing reliance on ground-based radar systems.

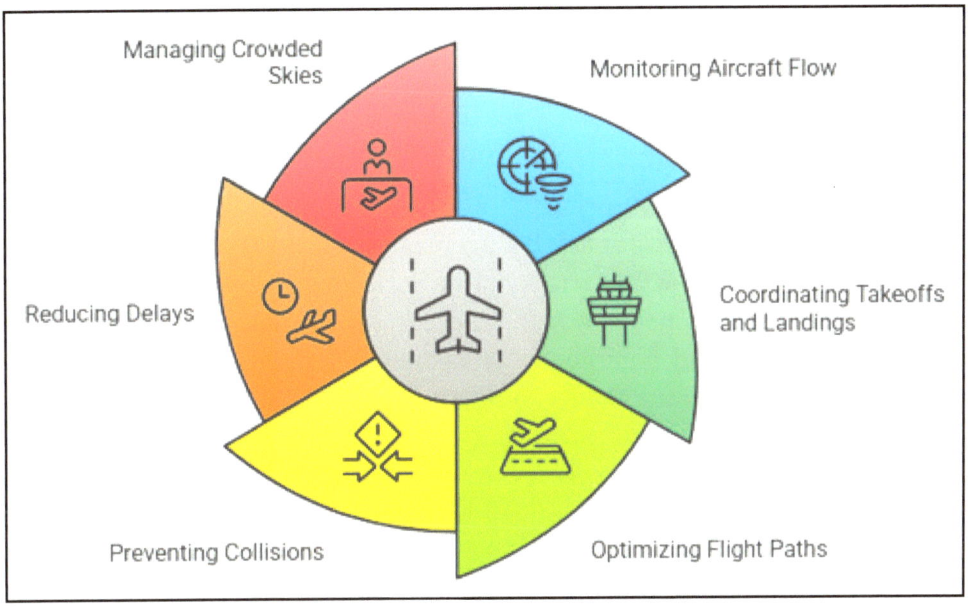

Ensuring Safe and Efficient Air Travel

The next phase of traffic control—known as Unmanned Traffic Management (UTM)—extends these principles to drones and small aerial systems. UTM platforms use cloud-based technologies to coordinate thousands of unmanned vehicles simultaneously, enabling safe low-altitude operations within urban areas.

Table: Evolution of Air Traffic Control Technologies

Generation	Technology Focus	Key Features
First (1940s–60s)	Radar-based ATC	Manual coordination, limited automation
Second (1970s–90s)	Computer-assisted ATC	Enhanced flight data processing and communication
Third (2000s–2020s)	Satellite and GPS-based systems	Real-time surveillance, global coverage
Fourth (Present–Future)	AI-driven and integrated UTM systems	

9.3.5 Integration of Digital Technologies

Digital transformation is redefining air traffic control and infrastructure management. Artificial intelligence assists in flight prediction, weather modeling, and decision support. Big data analytics optimize route planning, fuel consumption, and congestion control. Blockchain can enhance the security of data exchanges, ensuring the authenticity of flight records and maintenance logs. Cloud computing also plays a critical role by enabling distributed airspace management, where data from multiple sources—satellites, sensors, aircraft systems—are processed in real time. This creates a shared operational picture for pilots, controllers, and airport authorities, enhancing coordination and response.

9.3.6 Safety, Regulation, and Environmental Considerations

Safety remains the highest priority in airspace and traffic control systems. Regulatory frameworks, enforced by agencies such as the International Civil Aviation Organization (ICAO) and national authorities, define operational standards for manned and unmanned aircraft. These include communication protocols, altitude restrictions, and emergency procedures. Environmental considerations are increasingly integrated into infrastructure and airspace management. Efficient routing reduces emissions, while electrification of aircraft contributes to sustainability. Noise abatement zones, controlled flight paths, and green airport designs help balance urban development with ecological protection.

9.3.7 Elements for Sustainable Airspace Integration

The successful integration of infrastructure, airspace, and traffic control depends on a few key elements that define the resilience and adaptability of the system. critical elements include:

- Efficient infrastructure planning that anticipates future aviation needs.
- Integration of unmanned aerial systems (UAS) with conventional aviation.
- Deployment of smart sensors for continuous airspace surveillance.
- Use of renewable energy sources in airport and vertiport design.
- Global harmonization of airspace policies and regulatory standards.
- Adoption of AI and machine learning for predictive traffic management.
- Development of secure communication channels for data exchange.
- Real-time airspace visualization tools for operators and controllers.
- Investment in training and capacity-building for aviation professionals.
- Commitment to environmental sustainability and noise reduction.

These elements collectively build a resilient and efficient airspace system capable of supporting both traditional and next-generation aviation models.

9.4 SOCIETAL ACCEPTANCE AND NOISE FOOTPRINT

9.4.1 Interconnection between Noise and Public Perception

Noise pollution from aircraft operations has long been a critical environmental issue. Historically, the expansion of airports and increased air traffic have often faced resistance from nearby communities due to disturbances caused by aircraft noise. With the introduction of electric vertical takeoff and landing (eVTOL) aircraft and other forms of urban air mobility, this issue has taken on renewed significance. Although electric propulsion systems promise quieter operations, even modest levels of residual noise can influence how people perceive these aircraft. Societal acceptance is not determined solely by decibel levels but also by how individuals subjectively experience sound. Factors such as tonal quality, duration, frequency, and the context in which noise occurs all contribute to public reaction. For example, noise heard in rural or residential areas may provoke stronger negative responses than the same level of noise in industrial zones.

9.4.2 Psychology of Sound and Community Sensitivity

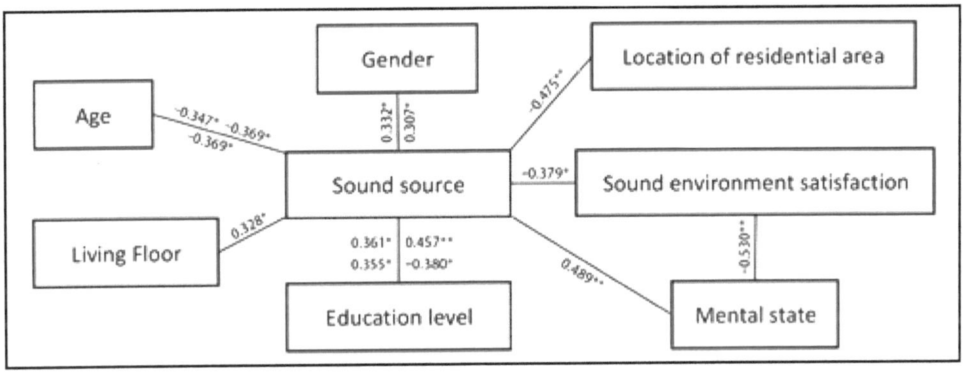

Soundscapes on Human Psychological Health

The human response to aircraft noise extends beyond measurable loudness to psychological and emotional factors. Research in environmental psychology has demonstrated that people's tolerance for noise depends on their attitudes toward the noise source. If an aircraft is perceived as beneficial—such as an emergency medical drone or public transportation service—communities may show higher tolerance. Conversely, if the same noise is associated with private or luxury travel, acceptance tends to decrease. Community sensitivity to noise is shaped by previous experiences, cultural attitudes toward technology, and trust in regulatory institutions. Furthermore, chronic exposure to noise can have measurable impacts on mental and physical health, including stress, sleep disturbance, and cardiovascular strain. These health implications intensify public opposition and complicate efforts to expand air mobility infrastructure.

9.4.3 Technological Innovations to Reduce the Noise Footprint

Engineers and designers have focused extensively on reducing the noise signature of modern aircraft. Electric propulsion systems, distributed electric propulsion (DEP), and advanced aerodynamic designs have been developed to address community noise concerns. In particular, the elimination of combustion engines reduces mechanical noise, while improved blade geometries minimize the acoustic disturbances generated by rotors and propellers. Noise abatement procedures—such as optimized flight paths, smoother ascent and descent profiles, and reduced hovering durations—also contribute to lowering noise exposure in sensitive areas. The use of acoustic modeling software enables manufacturers to predict how different designs will perform acoustically before production, allowing targeted noise mitigation strategies from the early stages of aircraft development. Despite these advances, the challenge remains complex, as urban environments amplify sound reflections, and multiple aircraft operating simultaneously may create cumulative noise effects.

9.4.4 Regulatory Frameworks and Standards

To foster societal acceptance, regulatory bodies worldwide have begun developing standards and certification procedures specific to noise emissions of new air mobility systems. Organizations such as the International Civil Aviation Organization (ICAO) and the Federal Aviation Administration (FAA) are updating guidelines to account for electric and hybrid aircraft. Urban air mobility introduces unique scenarios, such as vertical takeoff and landing in densely populated areas, which require entirely new assessment frameworks. Regulators are now considering metrics that combine acoustic measurements with human response data to better reflect real-world impacts. Public engagement in the regulatory process also plays a vital role. When communities feel involved in decision-making, trust increases, and resistance to new aviation technologies tends to decrease. Transparency in communication regarding expected noise levels, flight routes, and operational hours is therefore a key component of regulatory success.

9.4.5 Urban Planning and Noise Zoning

Effective integration of air mobility into cities requires strategic urban planning that accounts for noise footprints. Designing "vertiports" and flight corridors with community well-being in mind can significantly reduce conflict between operators and residents. For instance, flight paths can be routed along major highways, rivers, or industrial zones, where background noise levels are already higher, thereby masking aircraft sounds. Moreover, local zoning laws can restrict air operations during night hours or in proximity to hospitals and schools. Collaboration between aviation authorities, city planners, and acoustic engineers ensures that the benefits of air mobility—such as reduced traffic congestion and improved accessibility—are realized without compromising the livability of urban areas. A balanced approach that integrates technological, spatial, and social considerations can enhance both noise management and societal acceptance.

Table: Factors Influencing Societal Acceptance of Aviation Noise

Category	Description	Impact on Acceptance
Acoustic Factors	Loudness, tone, frequency, and duration of noise events	Direct physical discomfort and annoyance
Psychological Factors	Attitudes, trust, and perception of purpose	Influences tolerance and emotional response
Environmental Context	Urban density, background noise, and land use	Affects how noticeable noise is in daily life
Regulatory Transparency	Community engagement and communication	Builds trust and reduces opposition
Technological Innovation	Noise-reducing propulsion and design	Demonstrates industry commitment to sustainability

9.4.6 Socioeconomic and Cultural Dimensions

Societal acceptance of aviation noise is also shaped by socioeconomic and cultural dimensions. Wealthier communities often possess greater political influence to resist infrastructure projects that produce noise, while lower-income populations may bear disproportionate exposure due to fewer regulatory safeguards. Cultural factors further influence perceptions: in some societies, technological progress is embraced as a symbol of national pride, while in others, environmental preservation holds higher value. Effective communication strategies must therefore be tailored to local contexts, recognizing that a uniform policy cannot address all societal variations. Public education about the environmental advantages of electrified aviation, such as reduced greenhouse gas emissions, can also offset negative attitudes toward residual noise.

9.4.7 Building Public Trust through Engagement

Public trust is central to societal acceptance. Transparent communication, participatory planning, and responsiveness to community feedback can transform skepticism into support. Community workshops, simulation demonstrations, and virtual noise experience platforms allow residents to understand the expected impacts before aircraft deployment begins. Involving citizens in trial projects and data collection can foster a sense of shared ownership over urban mobility solutions. Governments and aviation companies that prioritize community well-being through social responsibility initiatives often encounter less resistance when introducing new technologies. Therefore, cultivating trust through continuous engagement is not a peripheral task—it is the foundation of sustainable aviation development.

9.5 PILOTLESS SYSTEMS AND AUTONOMOUS FLIGHT

Pilotless systems and autonomous flight promise transformative benefits in accessibility, efficiency, and new service models, but realizing that promise depends on coordinated advances across technology, regulation, infrastructure, and social license. Incremental steps that emphasize demonstrable safety, interoperable traffic management, and robust verification of AI components will pave the way to broader operations. Research into formal verification for learning systems, improved energy storage, and resilient communications will unlock more ambitious missions. Ultimately, the successful integration of pilotless aircraft will not be judged solely by technological prowess but by how well systems are governed, how transparently risks are managed, and how equitably the benefits are distributed across society.

9.5.1 Defining Pilotless Systems

Pilotless systems encompass a spectrum of architectures, from remotely piloted aircraft where a human operator controls one or many vehicles from a ground station, to supervisory autonomy where a human monitors and intervenes, when necessary, to fully autonomous aircraft that complete missions with no human-in-the-loop during normal operations. The terminology varies across stakeholders, with terms such as remotely piloted systems, uncrewed aircraft systems, and autonomous aircraft used in overlapping ways. What unites these systems is a reliance on onboard and offboard processing to handle perception, navigation, trajectory planning, and contingency management. Depending on the application, pilotless systems may be designed for short, repetitive missions in constrained environments or for long-range operations in mixed airspace alongside conventional crewed aircraft.

CHAPTER 10

Leading Projects and Case Studies

10.1 PIPISTREL VELIS ELECTRO

10.1.1 Technical Configuration and Design Features

The aircraft's design centers around an advanced electric powertrain system featuring a liquid-cooled electric motor specifically developed for aviation applications. This motor generates approximately seventy-seven horsepower for takeoff and maintains slightly lower power for continuous cruise operation. Energy storage comes from two separate lithium-ion battery packs connected in parallel, with a total capacity sufficient for typical training missions. The dual-battery architecture represents a crucial safety innovation, providing system redundancy that allows continued operation should one battery encounter issues. Strategic placement of these battery units maintains optimal aircraft balance, with one positioned in the nose section and another behind the cockpit area. The airframe incorporates a high-wing configuration constructed primarily from advanced composite materials, optimizing strength while minimizing weight.

10.1.2 Operational Performance Characteristics

Performance parameters align perfectly with the aircraft's primary mission as a training platform. The typical cruise speed reaches approximately eighty-eight knots, while the practical endurance allows for approximately fifty minutes of flight time plus required reserve margins. This endurance proves sufficient for standard training exercises including pattern work, maneuvers, and local area familiarization. The operational ceiling extends to twelve thousand feet, with climb performance meeting training requirements. Operational advantages include simplified starting procedures requiring no engine warm-up period, enabling rapid turnaround between sessions. The exceptionally quiet operation creates new possibilities for flight training near noise-sensitive communities where traditional aircraft face restrictions.

10.1.3 Primary Mission and Economic Advantages

The aircraft's fundamental purpose centers on revolutionizing pilot training through sustainable technology. Economic analysis reveals substantial cost advantages over conventional training aircraft, with direct operating costs demonstrating significant savings potential for flight schools. The primary cost components include aircraft depreciation, scheduled maintenance, and electrical charging expenses. The electrical energy consumption translates to remarkably low fuel costs compared to avgas or jet fuel. Maintenance requirements show notable reductions due to the simplified electric powertrain's reduced moving parts and elimination of traditional engine maintenance items. These economic benefits become increasingly substantial when considering fleet operations at training organizations, providing financial incentives alongside environmental advantages.

10.1.4 Global Implementation and Market Acceptance

International adoption continues expanding across multiple continents, with particularly strong presence in European markets supported by progressive regulatory frameworks and sustainability initiatives. Numerous flight training organizations have integrated these aircraft into their training fleets, accumulating substantial operational experience. The implementation strategy often involves phased adoption, allowing organizations to develop standardized procedures for electric aircraft operations. Military organizations have also shown interest, with several air forces acquiring aircraft for evaluation and pilot training purposes.

Planetary-Scale Implementation

10.2 ALICE BY EVIATION

Alice by Eviation marks a defining moment in aviation history. It showcases what is possible when technology, sustainability, and design converge toward a common goal: zero-emission air travel. Through its innovative structure, efficient propulsion, and quiet operation, Alice demonstrates that electric aviation is not a distant dream but a rapidly approaching reality.

While challenges remain, the progress achieved by Eviation underscores the momentum toward a cleaner and smarter aviation future. As global attention turns to sustainability, Alice stands ready to lead the next generation of regional flight—one powered not by fuel, but by electricity and human ingenuity. Eviation's long-term vision extends beyond Alice. As battery technology progresses and global infrastructure adapts, the company plans to expand its portfolio with aircraft capable of carrying more passengers and flying longer distances. Collaboration with energy firms, regulators, and airline operators will be essential to achieving large-scale deployment. The success of Alice could accelerate the electrification of short-haul aviation worldwide. Within the next decade, it is expected that electric and hybrid aircraft will dominate routes under 500 miles, transforming how passengers and cargo move regionally. Alice stands as a symbol of this transformation—an aircraft that merges innovation, sustainability, and practicality into a single design.

10.2.1 Concept and Development

Eviation Aircraft introduced Alice as part of a vision to decarbonize air travel and reduce the cost of operation through full electrification. Founded in 2015, the company sought to create an electric aircraft that could serve as a practical alternative to conventional regional jets and turboprops. Unlike traditional planes that rely on fossil fuels, Alice uses battery-powered electric motors to generate thrust, offering a sustainable and quiet flight experience.

The development of Alice involved partnerships with several global aerospace leaders. MagniX provided the electric propulsion system, while Daher and Honeywell contributed to avionics and system integration. The aircraft first took flight in September 2022, marking a milestone for the electric aviation sector. Its test flights demonstrated not only the feasibility of electric propulsion but also its potential to significantly reduce maintenance costs and carbon emissions.

10.2.2 Design and Technical Specifications

Alice features a sleek, futuristic design optimized for efficiency. The aircraft has a composite fuselage to minimize weight while maximizing aerodynamic performance. It is designed to carry nine passengers and two crew members, making it ideal for regional carriers, charter services, and corporate transport.

The aircraft's propulsion system is fully electric, powered by two Magni650 electric motors, each delivering up to 640 kW (858 hp). Its energy source—a lithium-ion battery pack—constitutes nearly 60% of the aircraft's total weight. Although battery energy density currently limits range, ongoing advancements in solid-state and high-density batteries are expected to extend its operational capabilities.

Alice's range is approximately 250 to 400 nautical miles, depending on payload and flight conditions, with a maximum cruise speed of 260 knots (482 km/h). Charging time is estimated at 30 minutes per flight hour, enabling fast turnaround for regional operations. The aircraft's operational altitude is designed around 10,000 feet, reducing pressurization requirements and improving energy efficiency.

Table: Specifications of Eviation Alice

Feature	Specification
Passenger Capacity	9 passengers + 2 crew
Propulsion	2 × Magni650 electric motors
Power Output	640 kW (each motor)
Maximum Range	250–400 nautical miles
Cruise Speed	260 knots (482 km/h)
Battery Type	Lithium-ion (advanced energy-dense cells)
Charging Time	30 minutes per flight hour
Maximum Takeoff Weight	16,500 pounds (7,484 kg)
Service Ceiling	10,000 feet
First Flight	September 2022

10.2.3 Sustainability and Environmental Impact

One of the most compelling features of Alice is its zero-emission design. By eliminating fuel combustion, Alice produces no carbon dioxide or nitrogen oxide emissions, significantly reducing aviation's environmental footprint. Traditional aircraft contribute nearly 2.5% of global CO_2 emissions, but electric aviation has the potential to bring that figure down dramatically.

Noise pollution is another area where Alice stands out. Its electric motors operate almost silently compared to jet or turboprop engines, making it ideal for urban and suburban airports where noise restrictions limit flight operations. This quiet performance could open new routes closer to population centers, supporting regional connectivity without disturbing communities.

Moreover, the aircraft's operational efficiency reduces maintenance needs. Electric motors have far fewer moving parts than conventional engines, minimizing wear, tear, and downtime. Combined with lower energy costs per flight hour, this makes Alice not only eco-friendly but also economically advantageous for operators.

10.2.4 Technological Innovations

Alice is not merely an electric version of existing aircraft—it is a complete rethinking of aircraft architecture. The integration of batteries, power electronics, and aerodynamics required extensive innovation. Eviation designed the fuselage and wings to maximize lift-to-drag ratio while housing large battery packs safely within the structure.

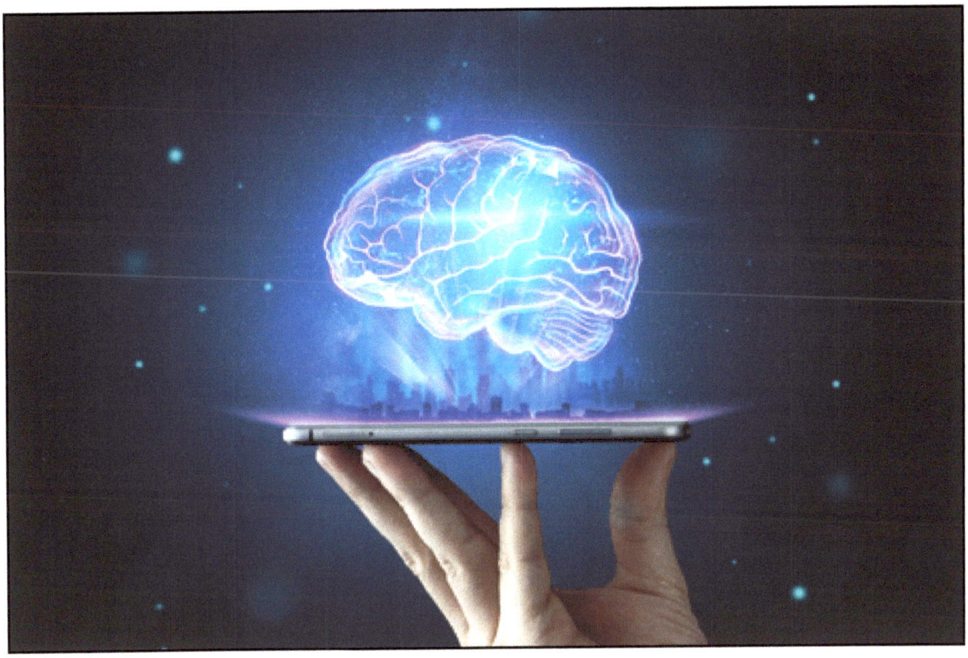

Technological Developments

The use of fly-by-wire technology allows for precise control with reduced mechanical complexity. Advanced thermal management systems ensure that the batteries maintain optimal temperature throughout flight, preventing overheating and enhancing safety. Eviation has also incorporated advanced avionics, including Honeywell's flight control systems, which provide intuitive pilot interfaces and enhance situational awareness. The aircraft's digital systems continuously monitor performance data, enabling predictive maintenance and ensuring reliability over its lifespan.

10.2.5 Operational Vision and Market Potential

Alice targets a growing market segment—short-haul regional transportation. With many airlines seeking to decarbonize their fleets and governments pushing for net-zero aviation goals, electric aircraft present a viable solution for routes under 500 miles. Regional carriers could use Alice to connect smaller cities, remote locations, and business hubs that are often underserved by larger jets. The aircraft's lower operating costs make it particularly appealing for on-demand air taxi services, corporate shuttle operations, and logistics providers handling lightweight cargo.

Eviation envisions Alice as the forerunner of an entire fleet of electric aircraft, inspiring future models with greater range and capacity as battery technology evolves. The aircraft has already attracted interest from several companies, including Cape Air and GlobalX, which have expressed intentions to incorporate Alice into their future operations.

10.2.6 Challenges and Limitations

Despite its promise, Alice faces several technological and regulatory challenges. Battery energy density remains the most significant barrier; current batteries store far less energy per kilogram than jet fuel, limiting range and payload capacity. Additionally, widespread adoption requires a network of charging infrastructure at airports—a system that is still in early stages of development. Certification also poses a challenge, as regulatory authorities like the FAA and EASA must establish new safety standards for electric propulsion systems. Maintenance procedures, emergency protocols, and thermal management all require thorough evaluation before electric aircraft can enter widespread service.

10.2.7 Features of the Eviation Alice

- Fully electric propulsion with zero carbon emissions.
- Lightweight composite structure for maximum aerodynamic efficiency.
- Nine-passenger configuration ideal for regional travel.
- Two Magni650 electric motors providing 1,280 kW combined power.
- Quiet operation suitable for urban airports and short routes.
- Fly-by-wire system with digital flight controls.
- Quick charging cycle supporting fast operational turnaround.
- Lower maintenance costs due to fewer moving parts.
- Reduced operating noise and environmental footprint.
- Designed as a scalable platform for future electric aircraft models.

These features position Alice as a benchmark for innovation in the electric aviation market, setting the standard for future developments.

10.2.8 Economic and Industrial Implications

The economic implications of Alice extend far beyond its manufacturer. The shift toward electric aviation could transform regional economies, reduce airline fuel expenses, and generate new opportunities in manufacturing, maintenance, and energy supply. Airports might develop charging hubs similar to electric vehicle stations, while renewable energy providers could partner with aviation companies to power electric fleets sustainably. The transition to electric flight also supports global goals of achieving carbon-neutral aviation by 2050, aligning with international agreements like the Paris Climate Accord. By pioneering commercial electric aircraft, Eviation not only challenges existing industry norms but also paves the way for new business models focused on sustainability and innovation.

10.3 JOBY AVIATION AND VERTICAL AEROSPACE (EVTOLS)

Joby Aviation and Vertical Aerospace stand at the forefront of a technological revolution that could redefine the future of mobility. Their efforts to design, certify, and deploy efficient, zero-emission aircraft represent more than just advances in aerospace engineering—they symbolize humanity's broader aspiration for sustainable innovation.

Through a combination of visionary leadership, cutting-edge technology, and strategic partnerships, these companies are laying the groundwork for a new era of flight that is cleaner, quieter, and more accessible. While challenges remain, the progress achieved thus far suggests that eVTOLs are not a distant dream but an imminent reality poised to transform urban life, reshape the aviation industry, and usher in a future where air travel becomes an integral part of daily transportation networks.

10.3.1 Vision and Foundation of Joby Aviation

Founded in 2009 by entrepreneur and engineer JoeBen Bevirt, Joby Aviation emerged from a vision to create environmentally sustainable air transport that could operate with minimal noise and zero emissions. The company is headquartered in Santa Cruz, California, and has been one of the most active and well-funded developers in the eVTOL sector.

Joby's central objective is to develop an aircraft capable of combining the speed of a small airplane with the vertical agility of a helicopter while maintaining quiet, safe, and cost-effective operation. The aircraft developed by Joby is designed to carry a pilot and four passengers, achieving a top speed of around 200 miles per hour with a range exceeding 150 miles on a single charge. This makes it suitable for both intra-urban and regional travel, positioning it as a viable alternative to congested ground transportation systems. Joby Aviation's design philosophy emphasizes simplicity, reliability, and efficiency.

The company's aircraft employs six tilting electric propellers that enable vertical takeoff and landing and transition seamlessly into forward flight. The distributed electric propulsion system not only enhances aerodynamic efficiency but also offers redundancy, increasing flight safety. Furthermore, the aircraft is engineered for low acoustic emissions, making it roughly 100 times quieter than conventional helicopters during takeoff and nearly silent in forward flight. These attributes address two of the primary concerns associated with urban air mobility—noise and safety—thereby supporting greater societal acceptance of eVTOL operations.

10.3.2 Vertical Aerospace: The UK's Answer to Sustainable Air Mobility

Founded in 2016 by Stephen Fitzpatrick, Vertical Aerospace is a British aerospace manufacturer headquartered in Bristol. Like Joby Aviation, the company's mission is to pioneer sustainable aviation solutions that reduce carbon emissions and urban congestion. Vertical Aerospace's flagship aircraft, the VX4, represents a culmination of years of research, collaboration, and advanced engineering.

The VX4 is designed to carry four passengers and one pilot, offering a range of up to 100 miles and a top speed of approximately 200 miles per hour. It operates entirely on electric power and is engineered to provide near-silent flight, ensuring compatibility with densely populated urban environments. Vertical Aerospace has positioned itself strategically through strong partnerships with major aerospace and technology firms.

Collaborations with companies such as Rolls-Royce, Honeywell, Leonardo, and Microsoft have accelerated the development of propulsion systems, avionics, and flight control technologies. Furthermore, Vertical Aerospace has received pre-orders from leading global airlines, including American Airlines, Virgin Atlantic, and Japan Airlines, reflecting significant commercial interest in its eVTOL solutions. These collaborations not only validate the company's technological credibility but also signal a major shift in the aviation industry's commitment to electrification and sustainability.

Table: Comparison of Joby Aviation and Vertical Aerospace

Feature	Joby Aviation	Vertical Aerospace
Founded	2009, California, USA	2016, Bristol, UK
Founder	JoeBen Bevirt	Stephen Fitzpatrick
Aircraft Model	Joby S4 eVTOL	VX4 eVTOL
Passenger Capacity	4 passengers + 1 pilot	4 passengers + 1 pilot
Range	~150 miles	~100 miles
Top Speed	~200 mph	~200 mph
Propulsion	6 tilting electric propellers	8 electric rotors
Partnerships	Toyota, NASA, U.S. Air Force	Rolls-Royce, Honeywell, Virgin Atlantic
Key Focus	Urban and regional air mobility	Commercial airline integration and urban mobility

10.3.3 Technological Innovations and Design Philosophy

Both Joby Aviation and Vertical Aerospace have approached eVTOL design with a focus on safety, simplicity, and operational efficiency. Joby's aircraft uses advanced carbon-fiber composite materials that provide a lightweight yet durable structure capable of withstanding high aerodynamic loads. The use of distributed electric propulsion allows for multiple smaller propellers rather than a single large rotor, significantly reducing mechanical complexity and noise output. Joby has also placed strong emphasis on battery optimization, seeking to balance power density with safety and longevity.

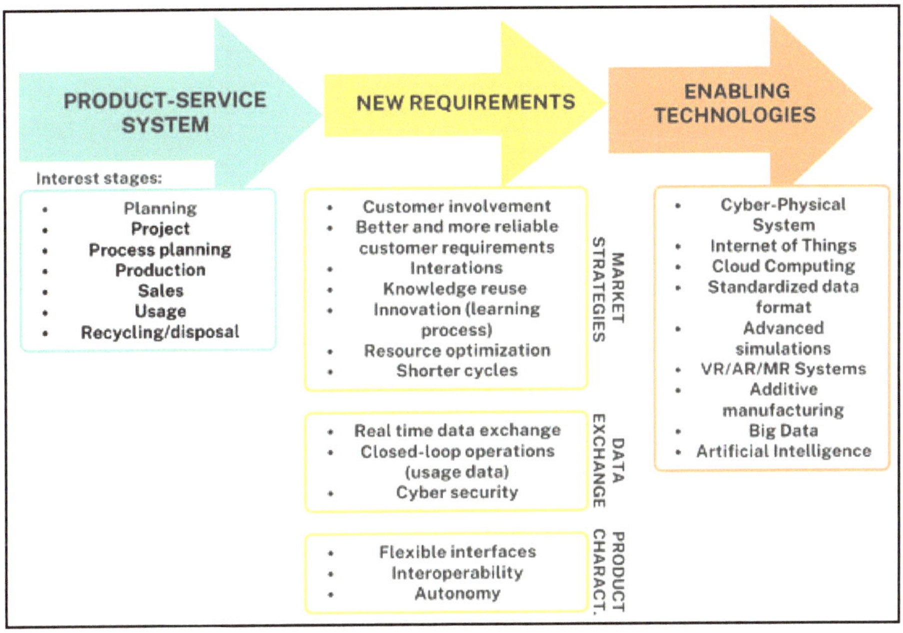

Technological Advancements and Design Approach

Similarly, Vertical Aerospace's VX4 is engineered for modularity and scalability. The company employs a quadruple redundant flight control system to ensure maximum reliability, even in the event of component failure. Its aerodynamic design minimizes drag while maintaining high lift efficiency during vertical and cruise flight modes. Vertical Aerospace has also adopted digital twin and simulation technologies to refine its aircraft's performance through virtual testing, thereby reducing physical prototyping costs and development time. Both companies are committed to adhering to stringent aviation safety standards, working closely with the Federal Aviation Administration (FAA) and the UK's Civil Aviation Authority (CAA) to obtain type certification.

10.3.4 Market Strategy and Commercial Deployment

Joby Aviation's approach to commercialization involves not only manufacturing aircraft but also operating them as part of an integrated aerial ride-sharing service. The company envisions a network similar to current ground-based ride-hailing systems, where passengers can book on-demand flights between cities or urban centers. In pursuit of this model, Joby has established partnerships with Uber

Technologies and has acquired Uber Elevate to integrate aerial mobility into future urban transport ecosystems. Additionally, Joby has entered into agreements with the U.S. Department of Defense and NASA to provide demonstration flights and gather operational data that support certification and future deployment. Vertical Aerospace, on the other hand, is focusing primarily on commercial partnerships with airlines and transport providers. Its goal is to supply eVTOL aircraft to established operators rather than running its own transport service. This approach allows Vertical Aerospace to scale production efficiently while relying on experienced aviation partners for flight operations and maintenance. By collaborating with established airline networks, the company aims to integrate its eVTOL technology into existing air travel systems, offering short-haul connectivity between airports and city centers.

10.3.5 Environmental and Economic Impact

The environmental benefits of Joby Aviation and Vertical Aerospace's eVTOL aircraft are profound. Both companies have prioritized all-electric propulsion to eliminate direct emissions and significantly reduce the carbon footprint of air travel. This commitment aligns with global climate goals and the aviation industry's broader push toward net-zero emissions by 2050. Moreover, the energy efficiency of electric motors compared to combustion engines ensures that operational costs remain low, making eVTOL transport potentially more affordable for consumers over time. From an economic perspective, eVTOLs have the potential to stimulate new industries, including battery manufacturing, vertiport construction, and digital air traffic management systems. The resulting ecosystem could create thousands of jobs and support the growth of sustainable urban economies.

10.3.6 Challenges and the Road Ahead

Despite remarkable progress, both Joby Aviation and Vertical Aerospace face substantial challenges before large-scale deployment can be realized. Regulatory approval remains a lengthy and complex process, as authorities must ensure that these aircraft meet rigorous safety and reliability standards. Battery technology continues to be a limiting factor, with current energy densities constraining range and payload capacity. Infrastructure development, including the establishment of vertiports, charging stations, and air traffic control systems tailored to low-altitude operations, is still in early stages. Additionally, societal acceptance will play a decisive role, as public concerns regarding safety, privacy, and noise must be effectively addressed.

10.4 LILIUM JET AND MULTIROTOR DESIGNS

Comparing the Lilium Jet concept and multirotor designs underscores a fundamental trade-off in eVTOL engineering between hover simplicity and cruise efficiency.

Winged, vectored-thrust aircraft pursue aerodynamic economy and higher-speed missions by introducing transition complexity and tighter integration of propulsion and flight control, while multirotors prioritize operational simplicity, hover performance, and modest infrastructure footprints at the cost of energy efficiency in cruise. Both approaches face shared constraints in battery energy, certification, noise mitigation, and public acceptance, and both will evolve as technologies and regulations mature. The coming decade will reveal which niches each design serves best and whether hybrid or convergent solutions will ultimately define routine urban and regional electric flight.

10.4.1 Lilium Jet: Concept and Distinctive Features

The Lilium Jet exemplifies a winged eVTOL concept that uses distributed electric propulsion combined with wing-borne cruise to maximize aerodynamic efficiency in forward flight. Rather than relying on large exposed rotors, Lilium's concept integrates many small, electrically driven ducted fans or jets into the wing and canard surfaces and vectors their thrust for vertical climb, transition, and efficient high-speed cruise. By carrying the aircraft's weight on aerodynamic lift during cruise, the design reduces power draw compared with an all-rotor vehicle, enabling longer range and higher cruise speeds for passenger missions. The trade-offs include added complexity during the vertical-to-horizontal transition, tight thermal and power-management demands, and the need to validate novel ducted-propulsion arrangements through extensive testing and certification.

10.4.2 Multirotor Architecture and Scaling

Multirotor designs—ranging from quadcopters to hexacopters and octocopters—adhere to a simpler aerodynamic principle: multiple rotors provide lift and control without reliance on fixed wings, making hover straightforward and control laws relatively mature thanks to decades of drone development. Their mechanical simplicity and incremental scalability have made multirotors the de facto solution for delivery drones and many prototype air taxis. However, scaling multirotors up for efficient, high-speed passenger transport is challenging because rotor-based lift must be sustained at cruise speeds where large disc areas and high power are required, producing poor energy efficiency and limiting practical range given current battery densities. Consequently, multirotors excel at short-range, low-speed missions but struggle to match the cruise efficiency of winged eVTOLs.

10.4.3 Aerodynamics and Transition Flight

Aerodynamic behavior during transition separates winged vectored-thrust aircraft and multirotors. A winged vehicle must manage a complex regime change: thrust vectoring produces vertical lift during takeoff and then progressively reorients to provide forward acceleration as control surfaces and wing lift take over. The transitional flight envelope is where sophisticated flight control, accurate sensing, and predictive modeling are essential to maintain stability and passenger comfort. Multirotors avoid a comparable transition complexity because their rotors generate lift for both hover and forward flight, but that lack of transition benefit comes at the cost of higher drag and induced power requirements in cruise, where rotor inflow and blade aerodynamics become increasingly inefficient relative to wing lift.

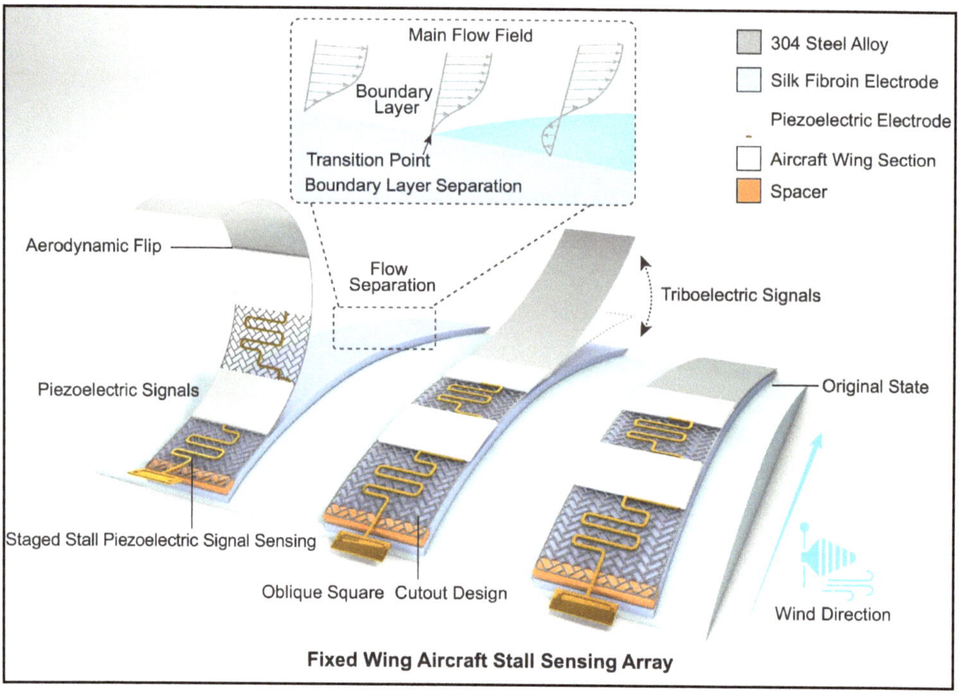

Digital Mapping of Surface Turbulence Status and Aerodynamic Stall on Wings of a Flying Aircraft

10.4.4 Propulsion Systems and Energy Considerations

Propulsion choices profoundly influence range, redundancy, and thermal considerations. Distributed electric propulsion offers redundancy and the potential for fine-grained control over lift distribution, but many small rotors or fans suffer from higher tip losses and can be less efficient than larger rotors at producing lift per unit power. Ducted fans, a hallmark of the Lilium approach, can mitigate some tip losses and offer shielding that affects noise and safety, but they introduce drag, mechanical complexity, and thermal-management needs for the densely packed motors and power electronics. Regardless of configuration, energy density of contemporary batteries constrains mission profiles; winged designs tend to leverage battery energy more efficiently in cruise, while multirotors accept shorter ranges in exchange for simpler operations and smaller landing sites.

10.4.5 Noise, Environmental and Community Impacts

Noise and environmental footprint are central to urban acceptance. Multirotors typically produce broadband acoustic signatures that can be perceived as intrusive when operations are frequent and close to populations, particularly because multiple rotors create complex interference patterns.

Ducted fans can alter noise spectra, potentially shifting energy to higher frequencies that carry differently through urban environments, and wing-borne cruise lowers power requirements over densely inhabited corridors, which can reduce sustained noise levels. Environmental impacts extend beyond acoustics: the lifecycle emissions of any eVTOL are tied to the electricity mix those charges batteries and to manufacturing impacts of batteries and composite airframes. Thoughtful routing, operational constraints, and noise-optimized propulsion design will be necessary to secure social license for routine operations.

10.4.6 Safety Architectures and Redundancy Strategies

Safety for both families depend on layered redundancy spanning hardware, software, and operations. Multirotors gain a degree of mechanical redundancy simply by having multiple rotors; designs can tolerate single-motor failures to varying extents provided control algorithms can reallocate thrust. Winged eVTOLs like the Lilium Jet rely on distributed fans for redundancy as well, but their safe outcome after failures also depends on wing-borne glide capability and the system's ability to manage controlled descent profiles. Both concepts must address battery thermal runaway, electromagnetic interference, and complex flight-control software certification. Achieving certifiable, deterministic behavior in the presence of faults is as much a systems-engineering challenge as it is an aerodynamic or propulsion one, and regulators are still developing frameworks tailored to these novel architectures.

10.4.7 Infrastructure and Operational Implications

Operational realities shape which designs are practical in urban contexts. Multirotors can operate from compact pads and mobile deployable sites, making them well suited to on-demand deliveries and operations where space is constrained. Winged eVTOLs, while capable of vertical landings, often demand more considered vertiport layout, approach corridors, and larger safety buffers due to higher cruise speeds and glide characteristics; their greater range and speed favor point-to-point regional routes and airport shuttles. Both concepts necessitate charging or battery-swap logistics, maintenance depots, and integrated airspace management systems that enable dense low-altitude traffic without compromising safety for conventional aviation. The interplay between vehicle design and vertiport economics will influence which models become commercially viable first.

10.5 AIRBUS, BOEING, AND TRADITIONAL OEM INITIATIVES

10.5.1 Evolution of Airbus Initiatives

Airbus has consistently advanced its agenda through innovative product development and integration of cutting-edge technologies into its commercial aircraft lines, notably the A320neo, A350 XWB, and the latest A321XLR. Focusing on fuel efficiency, lightweight composite materials, and highly digitalized cockpit and operational interfaces, Airbus' initiatives have often set industry standards for performance and operational costs. The company's incremental and disruptive innovation strategies are articulated through ambitious research and technology programs such as Clean Sky, a joint European public-private partnership aimed at developing next-generation environmentally friendly aviation technologies. Airbus has also played a crucial role in developing digital twins, advanced data analytics, and predictive maintenance platforms via its Skywise program, which aggregates flight, sensor, and operational data to enhance aircraft availability and optimize airline operations. These efforts are underpinned by a vast network of suppliers, global manufacturing sites, and investment in pilot training and customer support solutions, positioning the company as both a technology leader and service innovator.

10.5.2 Boeing's Strategic Focus and Product Development

Boeing has pursued a multifaceted approach to product innovation and market competitiveness, characterized by major investments in the 737 MAX, 787 Dreamliner, and ongoing efforts towards conceptualizing the New Midmarket Airplane (NMA). Boeing's strategy is anchored in advanced aerodynamics, propulsion technology, and strategic digitalization, seeking to enhance both aircraft performance and the airline customer experience. The company's digital transformation is embodied in the Boeing AnalytX suite, a comprehensive set of analytics tools designed to extract actionable insights from vast volumes of operational and maintenance data, supporting predictive maintenance, flight scheduling, and fuel optimization initiatives. Furthermore, Boeing's ecoDemonstrator program tests a wide range of emerging green aviation technologies and operational procedures onboard flying test beds, driving industry progress on fuel reduction, noise abatement, and emissions mitigation. Through partnerships with SAF producers, research organizations, and digital platforms, Boeing continues to pursue its sustainability commitments. The company's global footprint, relying on major partnerships in North America, Asia, and Europe, underpins a resilient supply chain that supports the simultaneous development of commercial, defense, and space systems.

10.5.3 Traditional OEM Initiatives and Competitive Positioning

Traditional OEMs in aerospace, including Embraer, Bombardier (now largely focused on business jets after selling its commercial aviation operations), ATR,

Mitsubishi Aircraft Corporation, and major Tier 1 suppliers such as Safran, Rolls-Royce, and Pratt & Whitney, contribute to the market with differentiated portfolios and targeted innovation. Embraer, for example, emphasizes regional connectivity and efficiency through the E2 family, promoting advanced aerodynamics and modernized avionics while expanding its services business model.

ATR, a leader in regional turboprops, invests in hybrid-electric propulsion research and operational efficiency for short-haul markets. Mitsubishi Aircraft Corporation, though challenged in commercializing its SpaceJet, still pursues regional jet design innovation. These traditional OEMs consistently innovate in propulsion, avionics, aerostructures, and support solutions to maintain relevance in a market characterized by the overwhelming presence of Airbus and Boeing. Tier 1 suppliers, meanwhile, drive competitive advantage through specialized component and subsystem advancements, ranging from additive manufacturing and new alloys to next-gen engine cores and software-defined aircraft systems.

10.5.4 Sustainability and Environmental Leadership

Leading in Sustainability and Environmental Responsibility

Airbus, Boeing, and traditional OEMs have intensified their sustainability initiatives in response to regulatory pressures, public expectations, and the urging of airline customers to decarbonize aviation. Airbus's ZEROe concepts publicly showcase its vision of hydrogen-powered commercial aircraft by 2035, exploring turboprop, turbofan, and blended wing body configurations that underscore the company's ambition to be first to market with zero-emission medium-haul aircraft. Airbus invests in sustainable aviation fuels (SAF), hybrid-electric propulsion, and circular economy approaches, aiming to improve the environmental profile of both production processes and aircraft operations.

Boeing likewise maintains aggressive targets for reducing lifecycle emissions, promoting broad SAF adoption, incorporating lightweight materials, and optimizing aerodynamics to reduce operational carbon intensity. The ecoDemonstrator platform, advancement of electric/hybrid-electric propulsion with partners, and continued commitment to research into next-generation sustainable designs keep Boeing at the forefront of the industry's energy transition. Among traditional OEMs and major suppliers, efforts are visible in lightweight engine and airframe designs, development of SAF-compatible engines, and increased focus on product lifecycle assessment to reduce overall environmental footprints.

10.5.5 Digitalization and the Role of Data

Digital transformation is central to the strategic initiatives of Airbus, Boeing, and traditional OEMs, reshaping manufacturing, aircraft performance monitoring, and customer engagement. Airbus's Skywise program, now established as an industry benchmark, enables operators to access real-time data, predictive analytics, and fleet management functionalities, fostering efficiency gains and reducing unplanned maintenance events.

Boeing's digitalization, similarly, encompasses aircraft health monitoring, predictive analytics via AnalytX, and efforts to digitalize the entire value chain from engineering through after-sales service. These approaches improve quality assurance, speed product development cycles, enhance operational safety, and create new value-added services for airlines. Traditional OEMs collaborate closely with customers and digital solution providers to integrate connected maintenance, e-enabled service platforms, and remote diagnostics, reflecting an industry-wide shift toward digital aviation ecosystems. Advanced manufacturing approaches such as additive manufacturing, automated assembly, and AI-driven supply chain management further reinforce the digital pivot among all aerospace OEMs.

10.5.6 Supply Chain Resilience and Global Partnerships

The complexity of modern aircraft production demands that Airbus, Boeing, and traditional OEMs manage extensive, multi-tiered global supply chains, increasingly focusing on resilience, flexibility, and risk mitigation in turbulent geopolitical and economic conditions. Both Airbus and Boeing have implemented advanced supplier quality management systems, invested in digital supply chain visibility, and fostered close relationships with logistics providers to ensure continuous production flow. The COVID-19 pandemic and recent global events highlighted vulnerabilities in the supply of semiconductors, raw materials, and critical subsystems, prompting a new wave of initiatives around onshoring, dual sourcing, and strategic buffer stocks. Partnerships with regional governments, R&D centers, and academic institutions further underpin the robustness of supply chains and access to next-generation talent and innovation.

10.5.7 Market Adaptation and R&D Investment

Market adaptation drives every major initiative of Airbus, Boeing, and traditional OEMs, compelling continuous investment in research & development (R&D), agile product roadmaps, and responsive services. Both Airbus and Boeing allocate substantial portions of annual revenues to R&D, ensuring the steady emergence of new aircraft configurations, enhanced avionic systems, and improved passenger experiences. As travel patterns shift, particularly with growing demand for narrowbody and regional aircraft post-pandemic, OEMs refocus development on versatile, low-cost, and environmentally efficient models. Traditional OEMs invest in technological differentiation, targeting emerging markets, new business models such as aircraft-as-a-service, and expansion into adjacent sectors like urban air mobility and advanced air cargo.

10.5.8 Collaborative Innovation and Industry Alliances

Industry alliances, joint ventures, and collaborative research platforms serve as critical accelerators for Airbus, Boeing, and traditional OEMs alike. Airbus works closely with global partners through international consortia, universities, and industry-academic alliances, fueling rapid concept validation and broad technology diffusion. Boeing collaborates with supply partners, regulators, and technology start-ups to share innovation risk and jointly develop solutions to common industry challenges, particularly in the areas of digitalization, sustainability, and safety. Traditional OEMs form alliances to share resources, scale their impact, and access new competencies, strengthening their roles in specialized market segments or emerging aviation applications. Alliances around SAF development, battery-electric research, and digital air traffic management illustrate the ecosystem-wide scale of cooperation required to address aviation's grand challenges.

10.6 NATIONAL PROGRAMS: NORWAY, USA, CHINA, INDIA

National programs in Norway, the USA, China, and India illustrate how diverse strategies can lead toward the same goal—a cleaner, smarter, and more connected world. Each country brings unique resources and expertise to the challenge of decarbonizing aviation and transport. Through innovation, collaboration, and policy support, these nations are paving the way for a sustainable future where electric aircraft and renewable energy redefine mobility. The progress achieved today marks the foundation of a new era in aviation—one where the skies remain open, quiet, and free from emissions.

10.6.1 Norway: The Pioneer of Electric Mobility

Norway has emerged as a global leader in the transition toward electric and sustainable transport. Its success stems from early policy interventions, public awareness, and infrastructure readiness. In aviation, Norway's government has declared its ambition to make all domestic flights electric by 2040, setting a powerful precedent for other countries to follow. The Avinor electric aviation program, initiated in collaboration with the Norwegian Civil Aviation Authority, aims to introduce short-haul electric flights across Norway's network of regional airports.

The geography of Norway, with its numerous fjords and short domestic routes, makes it an ideal environment for electric aviation. The partnership between Avinor and aircraft developers like Eviation and Heart Aerospace further accelerates the country's path toward zero-emission flight. Norway's extensive use of renewable hydropower ensures that the electricity used to charge electric aircraft and vehicles is also green, resulting in a nearly zero-carbon transportation system. Additionally, Norway's experience in electric cars—where over 80% of new vehicles sold in 2024 were electric—provides valuable lessons for aviation electrification. Government incentives, tax exemptions, and public trust have positioned Norway as a living example of a nation powered by clean energy innovation.

10.6.2 United States: Innovation through Collaboration

The United States has a long history of leading in aerospace innovation, and its national programs reflect a strong commitment to transforming the aviation landscape through advanced technologies. Several major initiatives have been launched to promote electric aviation, sustainable fuels, and modernization of airspace management systems.

The Federal Aviation Administration (FAA) and NASA play central roles in coordinating efforts among public and private stakeholders. NASA's Electric Powertrain Flight Demonstration (EPFD) program supports the testing of hybrid-electric and fully electric propulsion systems. Companies like Eviation, Joby Aviation, Archer Aviation, and Beta Technologies have received both federal and state-level support to develop next-generation electric aircraft for regional and urban transport. Another important initiative is the Sustainable Flight National Partnership, which focuses on improving aircraft efficiency, integrating renewable fuels, and supporting advanced air mobility (AAM). The U.S. Department of Energy (DOE) is also investing in high-performance batteries and hydrogen fuel technologies for aviation applications. At the state level, programs in California, Washington, and Texas are supporting charging infrastructure, vertiport development, and urban air mobility corridors. This combination of research, industry partnership, and regulatory adaptation demonstrates America's comprehensive approach to creating a sustainable and commercially viable electric aviation ecosystem.

10.6.3 China: Scaling Up Electric and Hybrid Aviation

China's approach to electric aviation and sustainable transport is characterized by scale, speed, and centralized planning. The Chinese government has integrated green aviation into its national strategies such as "Made in China 2025" and the 14th Five-Year Plan, which emphasize energy efficiency and technological self-reliance. China's Civil Aviation Administration (CAAC) oversees multiple programs supporting electric, hybrid, and hydrogen-powered aircraft. Domestic manufacturers such as COMAC, Aviation Industry Corporation of China (AVIC), and TCab Tech are developing electric vertical take-off and landing (eVTOL) aircraft and regional electric planes. These projects are supported by major state-owned enterprises and energy companies, ensuring strong financial and technical backing.

In addition to aircraft development, China has prioritized the establishment of a robust charging and maintenance infrastructure for electric aviation. Pilot projects in cities like Shenzhen, Shanghai, and Guangzhou are testing eVTOL operations for urban logistics and passenger transport. The nation's vast manufacturing base allows rapid scaling of electric motors, batteries, and aviation components, giving China a significant advantage in cost efficiency. Moreover, China's integration of artificial intelligence (AI) and 5G networks into air traffic management systems enables safe coordination of unmanned and manned electric aircraft. This high-tech infrastructure supports the government's long-term vision of a fully integrated, sustainable air transport network by 2050.

10.6.4 India: Building Sustainable Aviation Ecosystems

India's aviation and transport sectors are rapidly expanding, and with that growth comes the challenge of minimizing environmental impact. The Indian government has launched multiple programs focused on sustainability, renewable energy, and electric mobility to reduce the country's carbon footprint. In aviation, India's Ministry of Civil Aviation (MoCA) and DGCA (Directorate General of Civil Aviation) are exploring pathways for electric and hybrid aircraft deployment under the National Civil Aviation Policy (NCAP) framework.

Research collaborations between Hindustan Aeronautics Limited (HAL), Council of Scientific and Industrial Research (CSIR), and start-ups are underway to develop indigenous electric propulsion technologies. On a broader scale, India's National Electric Mobility Mission Plan (NEMMP) and FAME (Faster Adoption and Manufacturing of Hybrid and Electric Vehicles) initiatives lay the foundation for large-scale electrification of transport systems. The success of electric vehicle adoption has spurred discussions on creating similar infrastructure for small aircraft and urban air taxis. Furthermore, India's strong focus on renewable energy—particularly solar power—aligns perfectly with the needs of electric aviation. Solar-powered airports like Cochin International Airport, the world's first airport fully powered by solar energy, demonstrate India's innovative approach to sustainable infrastructure. These developments signal a future where clean aviation could complement the country's renewable energy transition.

Table: National Programs

Country	Key Initiative	Primary Focus	Goal Year
Norway	Avinor Electric Aviation Program	Domestic electric flights	2040
USA	NASA EPFD & Sustainable Flight Partnership	Electric propulsion, AAM	2035
China	CAAC Green Aviation Plan	Large-scale electric & hybrid aircraft	2050
India	NCAP & FAME Schemes	Electric mobility, regional aviation	2040

10.6.5 Common Goals and Divergent Strategies

Although these nations differ in their technological approaches and policy frameworks, their goals converge on three primary fronts—reducing emissions, achieving energy efficiency, and enhancing connectivity. Norway emphasizes renewable energy integration, the USA focuses on innovation and collaboration, China excels in large-scale implementation, and India aims to combine sustainability with affordability. Each country tailors its programs to its strengths: Norway's hydroelectric capacity, America's aerospace expertise, China's manufacturing scale, and India's renewable energy ambitions. Together, they represent the global blueprint for sustainable aviation and transportation.

CHAPTER 11

Challenges and Barriers

11.1 BATTERY LIMITATIONS: RANGE, WEIGHT, AND COST

The limitations of current battery technology—restricted range, excessive weight, and high cost—remain the defining challenges of electric aviation. Yet, these very obstacles drive innovation, pushing scientists and engineers to reimagine energy storage for flight. The convergence of technological progress, economic investment, and environmental necessity will ultimately determine how quickly electric aircraft become a reality. Overcoming these barriers will not only transform aviation but also contribute to the broader global transition toward clean, efficient, and sustainable transportation systems.

11.1.1 Crucial Role of Batteries in Aviation

In electric aviation, batteries are not merely components—they are the central determinants of aircraft capability. Unlike liquid fuels that contain vast amounts of energy per unit mass, batteries store energy electrochemically, which inherently limits their energy density. This means that even the most advanced lithium-ion batteries cannot yet match the energy storage efficiency of traditional aviation fuels such as jet kerosene. As a result, electric aircraft have shorter flight ranges and smaller payload capacities compared to their fuel-powered counterparts. The balance between performance, safety, and operational cost is intricately tied to battery characteristics, and improvements in this field are critical to enabling viable electric flight at commercial scales.

11.1.2 Range Constraints and Energy Density Challenges

The range of an electric aircraft is primarily determined by the specific energy of its battery, measured in watt-hours per kilogram (Wh/kg). Traditional jet fuel provides approximately 12,000 Wh/kg, whereas state-of-the-art lithium-ion batteries offer around 250–300 Wh/kg. This vast difference in energy density means that electric aircraft require far more mass in batteries to achieve comparable flight durations. For short-haul and urban air mobility applications, such as electric vertical takeoff and landing (eVTOL) vehicles, current battery performance is sufficient to provide ranges of 50 to 150 miles. However, for regional and long-haul aviation, these energy densities fall significantly short. The heavier the battery pack, the greater the energy required to lift and propel the aircraft, creating a self-limiting cycle that restricts achievable distance. To address range limitations, researchers are developing next-generation batteries, including solid-state, lithium-sulfur, and lithium-air chemistries, which promise higher energy densities. Solid-state batteries replace the liquid electrolyte with a solid one, offering improved safety and potentially higher capacity. Lithium-sulfur batteries can theoretically achieve energy densities two to three times greater than conventional lithium-ion, while lithium-air technology could rival the energy content of fossil fuels. Yet, these technologies remain largely experimental, with challenges such as short life cycles, instability, and manufacturing scalability delaying their commercial implementation. Until such technologies mature, range will remain a fundamental limitation for electric aviation.

11.1.3 Weight Paradox in Battery Design

Weight represents one of the most formidable constraints in electric aircraft engineering. Every additional kilogram of battery increases the aircraft's total mass, requiring more lift and consequently more energy to sustain flight. This paradox creates a delicate trade-off between battery capacity and aircraft efficiency. Unlike automobiles, which can afford to carry heavy battery packs without catastrophic effects on performance, aircraft are highly sensitive to weight variations. The energy needed for takeoff, climb, and cruise must all come from the same stored energy source, meaning that excessive battery mass can drastically reduce payload capacity and operational range.

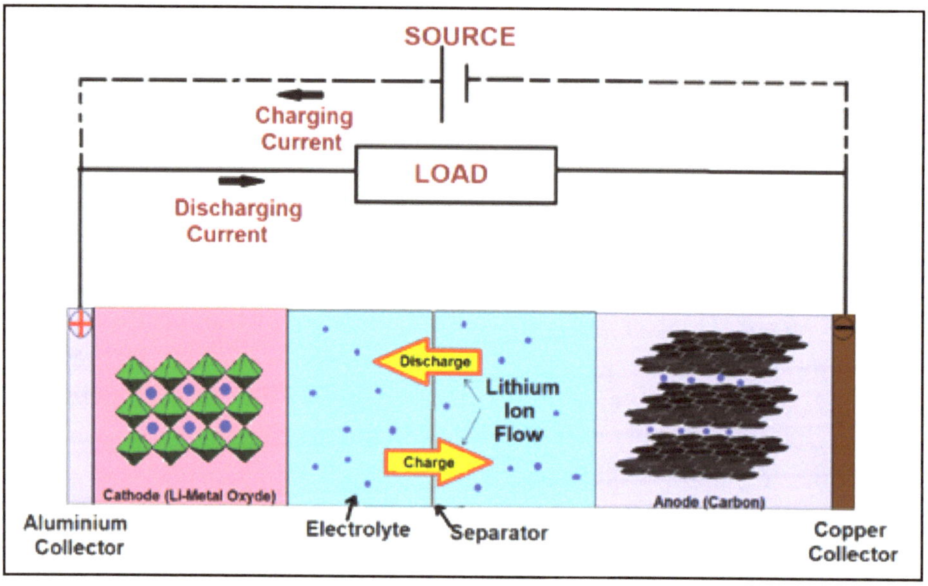

Battery Mass Paradox in Design Considerations

Designers attempt to mitigate this challenge through lightweight materials, aerodynamic optimization, and distributed propulsion systems that improve lift-to-drag ratios. Carbon-fiber composites, for instance, are used to construct airframes that minimize structural weight. However, these improvements cannot fully compensate for the relatively low energy density of batteries. A practical example can be seen in small eVTOL prototypes that allocate nearly one-third of their total mass to batteries, leaving limited capacity for passengers and cargo. The energy requirements for vertical takeoff and hover flight are especially demanding, as these operations consume more power than horizontal cruise. Consequently, the relationship between weight and performance defines the fundamental engineering boundary for electric aircraft design.

11.1.4 Cost Barriers in Battery Production and Operation

The cost of battery production and maintenance significantly influences the economic viability of electric aircraft. Lithium-ion batteries, currently the industry standard, are expensive to manufacture due to the high cost of raw materials such as lithium, cobalt, and nickel.

These materials are not only limited in supply but also subject to geopolitical and environmental concerns associated with mining and processing. The price of batteries is measured in dollars per kilowatt-hour (kWh), and although costs have declined dramatically over the past decade—from around $1,000 per kWh in 2010 to under $150 per kWh today—they remain a substantial expense for aviation applications, where larger capacities and higher safety standards are required.

In addition to production costs, maintenance and replacement represent ongoing financial challenges. Battery degradation over time reduces performance and flight range, necessitating periodic replacement to maintain safety and reliability standards. Unlike aircraft engines, which can operate for thousands of hours before overhaul, batteries have a finite cycle life, typically limited to a few hundred charge and discharge cycles. This short lifespan increases operating costs and reduces profitability, especially for commercial operators. Moreover, end-of-life disposal and recycling of batteries introduce environmental costs that must be factored into overall sustainability assessments.

Table: Comparison of Key Battery Limitations in Aviation

Limitation	Description	Impact on Aircraft Performance
Range	Limited by low energy density of current battery chemistries	Restricts flight distance and endurance
Weight	High battery mass reduces payload and increases energy demand	Limits efficiency and operational capability
Cost	High material and production costs with short cycle life	Affects economic feasibility and scalability

11.1.5 Technological Efforts Toward Improvement

Despite the current constraints, the global research community is actively pursuing innovations to overcome battery limitations. Advances in nanomaterials, electrode design, and energy management systems are improving performance incrementally. For instance, high-nickel cathodes increase storage capacity while reducing dependence on cobalt, which is both expensive and ethically problematic. Silicon anodes are being explored to replace traditional graphite, offering up to ten times the theoretical energy capacity. Additionally, advanced thermal management systems are being designed to enhance battery safety by preventing overheating during high-power operations such as takeoff and climb.

Battery management systems (BMS) also play a critical role by optimizing charge distribution and monitoring health status to extend lifespan. Researchers are investigating hybrid energy architectures that combine batteries with fuel cells or supercapacitors to balance power delivery and endurance. These hybrid systems may bridge the gap between current limitations and the energy demands of long-range electric flight. The transition from laboratory prototypes to scalable, certifiable aviation-grade batteries, however, remains a complex and resource-intensive process that requires close collaboration between aerospace manufacturers, material scientists, and regulatory agencies.

11.1.6 Economic and Environmental Implications

The economic implications of battery limitations extend beyond production costs to infrastructure investment and operational models. Airports and vertiports must develop new charging infrastructure capable of supporting rapid turnaround times between flights. Energy storage and grid integration will also become crucial as fleets of electric aircraft draw significant electrical power. While battery-based propulsion eliminates direct carbon emissions during flight, the environmental footprint of battery production and recycling cannot be ignored. Sustainable sourcing of lithium and other materials, as well as closed-loop recycling systems, are necessary to ensure that the environmental benefits of electric aviation are not offset by upstream ecological damage.

Societal perspective, addressing these limitations is key to achieving public acceptance of electric aviation. The promise of quieter, cleaner flight appeals to urban communities, but cost and accessibility will determine whether electric air mobility can become a mainstream mode of transportation or remain a niche innovation. Reducing battery costs through technological innovation and economies of scale will be essential to achieving commercial viability.

11.2 CHARGING INFRASTRUCTURE AND GRID CONSTRAINTS

Meeting the dual goals of widescale electrified mobility and a resilient, low-carbon grid requires integrated planning across utilities, regulators, vehicle operators, and urban planners. Short-term actions include targeted pilots that pair high-power charging with co-located storage, tariff experimentation to incentivize managed charging, and streamlined permitting to reduce lead times. Medium- and long-term solutions involve distribution system upgrades where economically justified, continued investment in forecasting and control platforms, and regulatory evolution to recognize distributed flexibility as a grid resource. Ultimately, success depends on aligning technical solutions with market incentives and public policy so that charging infrastructure expands equitably without creating structural constraints on the electricity system.

11.2.1 Types of Charging and Power Requirements

Charging installations range from low-power residential outlets to high-power fast-charging stations and depot infrastructures for fleet operations. Residential charging often uses lower power levels suited to overnight replenishment while public and corridor chargers provide rapid replenishment at higher power. Fleet depots and vertiports present different technical requirements, with depot chargers designed for scheduled bulk charging and potential battery-swapping solutions, and vertiports for electric aircraft likely demanding very high instantaneous power, substantial thermal management, and fast turnaround rates. Each category places distinct demands on transformers, cabling, and site electrical infrastructure, and higher power solutions require careful attention to cable sizing, connector cooling, and safety systems to manage heat and mechanical wear.

11.2.2 Grid Challenges and Constraints

Electrical distribution systems were not originally designed for synchronized, high-density vehicle charging, and they face several practical constraints. Local transformers and feeder cables have thermal limits that govern how much sustained power can be drawn without upgrades, while voltage regulation challenges arise when large, variable loads cluster on a given circuit. Protection coordination and fault current capabilities must be reassessed when new loads change the electrical behavior of a neighborhood. Upgrading substations and feeders can be costly and time-consuming, often requiring trenching and coordination with multiple stakeholders. Moreover, rapid charging and high-power installations can increase harmonic distortion and electromagnetic interference, requiring power-conditioning equipment and robust design practices.

11.2.3 Forecasting, Clustering Effects, and Peak Loads

Forecasting charging demand is inherently uncertain because it combines human mobility patterns, adoption curves, and technological shifts. Clustering of chargers at workplaces, retail centers, transit hubs, and vertiports can produce localized peaks that are far higher than aggregate city-level estimates would suggest. Time-of-day behavior, seasonal load variations, and fleet charging schedules complicate planning and can compound existing peaks from heating or cooling. Accurate forecasting therefore requires granular, scenario-based modeling that considers spatial-temporal correlations, behavioral elasticity in response to pricing signals, and potential technology shifts such as the proliferation of fast chargers or battery swapping in commercial fleets.

11.2.4 Smart Charging, Demand Management, and Market Mechanisms

Managed charging strategies can substantially mitigate grid impacts by shifting or modulating power to match system capabilities and price signals. Smart charging systems enable centralized or distributed control that reschedules charging windows, throttles power during local congestion, and participates in grid services. Dynamic tariffing and time-of-use rates provide economic incentives for load shifting, while aggregators can pool distributed load flexibility to bid into ancillary service markets. Vehicle-to-grid capabilities extend the potential by allowing aggregated fleets to provide short-term capacity, frequency response, and peak shaving, although the viability of V2G depends on battery lifecycle considerations, contractual arrangements, and regulatory frameworks that recognize distributed assets as grid resources.

11.2.5 Local Energy Solutions and Resilience

Co-locating energy storage and on-site generation with charging facilities offers a practical pathway to reduce grid upgrade costs and enhance resilience. Battery energy storage systems can act as buffer resources that smooth demand spikes, provide fast response for grid services, and store locally generated renewable energy to offset peak draws. Microgrids that combine solar generation, storage, and controllable loads can operate in grid-connected and islanded modes to maintain critical services during outages. For high-power facilities such as fleet depots and vertiports, hybrid designs that blend grid supply with storage and curtailed renewable output can minimize demand charges, reduce the need for immediate distribution upgrades, and improve operational reliability.

11.2.6 Siting, Permitting, and Urban Integration

Practical deployment hinges on thoughtful siting and streamlined permitting. Urban deployment often requires curbside accommodation, retrofits to existing parking facilities, and considerations for accessibility and pedestrian safety. Charging sites must balance proximity to demand centers with electrical feasibility; sometimes the optimal electrical connection is not at the highest-demand location, which necessitates last-mile solutions like short-distance feeder upgrades or microgrid links. Regulatory and zoning processes must evolve to speed installations while ensuring safety and aesthetic concerns are addressed. For vertiports and large depots, land use planning must incorporate not only electrical capacity but also staging areas, thermal management infrastructure, and traffic flow to and from the site.

11.2.7 Economics, Business Models, and Policy

The economics of charging infrastructure are shaped by capital costs, operational expenses, and electricity tariff structures, particularly demand charges that penalize short-term high usage. Business models vary from utility-owned infrastructure to third-party charging operators and fleet-managed systems, and public-private partnerships often accelerate deployment in underserved areas. Policy interventions such as grants, low-interest financing, and tariff reforms that decouple fixed-grid costs from volatile demand peaks can catalyze investment. Equitable access requires deliberate subsidization or regulatory mandates to avoid concentration of infrastructure only in affluent areas, and incentives for co-located storage and renewable generation can improve both environmental outcomes and grid compatibility.

11.2.8 Standards, Interoperability, and Cybersecurity

Operational Interoperability and Cyber Defense

Interoperability across charging hardware and software is critical for usability and efficient grid interaction; standardized communication protocols and payment systems reduce friction for users and streamline operational coordination. Secure communication channels ensure that managed charging commands, telemetry, and market participation cannot be tampered with, and robust authentication frameworks protect against unauthorized access. Cybersecurity and supply-chain integrity are particularly important where charging infrastructure participates in grid operations or where large numbers of distributed assets could be commandeered to create synchronized disruptions. Standards development and certification programs must progress in parallel with deployment to preserve safety and resilience.

11.3 REGULATORY AND CERTIFICATION BOTTLENECKS

11.3.1 Nature of Regulatory Oversight and Compliance

The regulatory environment in aviation is shaped by intricate documentation, strict traceability, and extensive audit requirements, all of which are fundamental to maintaining safety, airworthiness, and operational standards. Authorities such as the FAA and EASA demand comprehensive record-keeping from the manufacturing stage through the entire service cycle, including every part, subsystem, and maintenance procedure.

Meeting these requirements necessitates seamless digital and physical data trails, and any breakdown in these processes can trigger audits, operational delays, or even the grounding of aircraft. The documentation burden, coupled with the layered supplier ecosystem spanning Tier-1, Tier-2, and lower-tier components, creates a situation where misalignments or communication failures can result in production bottlenecks, excess inventory, or non-compliance penalties, ultimately affecting both the supply chain and the manufacturer's timelines for bringing new products to market.

11.3.2 Certification Delays and Industry Impact

The pace of aircraft certification has slowed considerably as regulatory agencies grapple with the inadequacies of legacy rules, especially those written for older mechanical systems that do not readily accommodate advanced avionics, automation, fly-by-wire technologies, or electric and hybrid propulsion systems. This mismatch forces manufacturers to seek special provisions and exceptions, a process that involves extensive negotiation, additional documentation, and often years of delay before approval is granted for new aircraft designs or system upgrades. Despite the deployment of high-fidelity simulations and digital twins capable of validating performance and safety, current regulations generally prioritize physical flight testing and exhaustive manual verification, which adds redundancy and inefficiency to the process. The FAA's planned revamp of speed and performance certification rules aims to address this by transitioning toward more performance-based standards that accept analytical modeling as primary evidence, potentially reducing the reliance on repetitive flight tests and accelerating approval timelines if successful.

11.3.3 Financial and Operational Ramifications

Regulatory bottlenecks impose considerable financial and operational risks on manufacturers, airlines, and suppliers alike. Delays in certification and compliance not only contribute to increased R&D and production costs, but they also disrupt fleet planning, increase training expenses, and lead to missed market opportunities. For instance, delayed mandates such as the FAA's secondary cockpit barrier requirement have led to contentious debate regarding their cost-effectiveness, industry feasibility, and practical implementation schedules. Aerospace giants and regional airlines must navigate retrofitting costs and training delays, while pilot unions advocate for urgent compliance due to persistent security threats. In this climate, regulatory uncertainty can further complicate investor assessments of aerospace profitability and drive-up airline compliance costs, particularly for smaller carriers lacking scale advantages.

11.3.4 System Integration and Legacy Aircraft

The complexity of certifying modern aircraft systems is exacerbated by the integration of increasingly sophisticated subsystems and the ongoing need to retrofit legacy fleets. Certifying upgrades, such as advanced avionics or new winglets for older models, requires meticulous safety and compatibility assessments and often exposes the limitations of existing certification processes. Legacy aircraft present additional challenges due to the need for compatibility with current standards, and the intricacies of systems integration require not only technical validation but also comprehensive training for engineers, pilots, and maintenance staff. The lessons learned from recent incidents, such as the Boeing 737 MAX, have reinforced the importance of certified crew resource management, highlighting the necessity of robust protocols and documentation for training, communication, and operational coordination.

11.3.5 Fragmented Compliance and Workflow Gaps

Operational bottlenecks frequently arise from fragmented compliance systems, where inspection records, task statuses, and regulatory documentation are managed across disconnected platforms or manual spreadsheets. This fragmentation facilitates compliance gaps, missed inspection intervals, or overlooked directives that can result in grounded aircraft, failed audits, or safety risks.

Airlines and MROs are increasingly adopting integrated, centralized digital platforms to unify compliance tracking and maintenance workflows, which promises to reduce errors, improve audit readiness, and maintain asset airworthiness. However, the transition from legacy systems to modern, connected solutions can itself introduce complexity and temporary risk, particularly during periods of heightened regulatory scrutiny.

11.3.6 Supply Chain Complexity and Regulatory Strain

The aviation supply chain, comprised of a global network of tiered suppliers, is highly sensitive to regulatory changes and certification requirements. Delays in component approvals, material shortages, and vendor hold-ups can cascade across production schedules and maintenance operations. The management of aging fleets and the diversity of aircraft models further complicate procurement and inventory control, especially when new regulatory mandates require immediate retrofits or modifications that suppliers are not equipped to deliver at scale. This can trigger shortages, operational downtime, and increased costs, amplifying the regulatory burden across the entire supplier ecosystem.

11.3.7 Sustainability and Decarbonization Hurdles

Environmental regulation, notably surrounding the adoption of sustainable aviation fuels (SAF) and decarbonization measures, presents an additional layer of regulatory bottleneck, with inconsistent regional policies, unclear standards, and slow alignment of sustainability mandates. Manufacturers and airlines face uncertainty in their planning due to patchy policy frameworks, shifting subsidy timelines, and differing feedstock advantages across geographic markets. The lack of harmonized global rules slows down investment in SAF infrastructure and technology, demanding clearer, more consistent regulatory guidance to enable industry-wide adoption and capitalize on environmental progress.

11.3.8 Slow Regulation and Innovation Stagnation

Industry observers increasingly argue that slow regulation creates a pause for innovation, particularly for new aircraft types like electric planes, autonomous drones, and advanced air traffic management systems. The inability of regulatory bodies to keep pace with technology means that even proven innovations can remain stalled in approval limbo, blocking widespread adoption and preventing the realization of benefits such as reduced emissions, enhanced safety, and improved operational efficiency. Air traffic controllers, airlines, and airports are similarly affected by outdated regulatory requirements that mandate continued use of legacy systems, despite the availability of more advanced tools and procedures. The resulting cycle holds back industry progress, with older systems remaining in use longer than necessary and the promise of safer, greener, and more reliable air travel delayed by regulatory inertia.

11.3.9 Prospects for Reform and Modernization

Momentum for regulatory reform is building, with agencies recognizing the urgent need to evolve certification frameworks to accommodate new technologies and streamline compliance. The FAA's emphasis on performance-based validation, greater reliance on analytical modeling, and the adoption of digital platforms across the industry represent significant steps toward more responsive, science-driven regulation. Stakeholders hope these changes will minimize bespoke negotiation, eliminate redundant flight testing, and create a standardized, predictable certification pathway that accelerates technology adoption while maintaining safety. As the global market for commercial and sustainable aircraft continues to expand, the evolution of regulatory processes will play a defining role in shaping the speed and direction of aviation innovation for years to come.

11.4 PUBLIC PERCEPTION AND SAFETY CONCERNS

Public perception and safety concerns are pivotal in shaping the future of aviation. The success of electric and autonomous aircraft depends as much on public confidence as on engineering excellence. Addressing fears through education, regulation, and transparent communication ensures that innovation aligns with societal trust. By combining rigorous safety standards, open dialogue, and visible performance reliability, the aviation industry can transform skepticism into confidence. In doing so, it not only advances technology but also redefines what safety means in the modern era—an era where innovation and trust must coexist for the skies of tomorrow to truly open.

11.4.1 Evolution of Public Perception in Aviation

Historically, the acceptance of new aviation technologies has followed a predictable pattern—initial skepticism, gradual acceptance, and eventual normalization. In the early 20th century, air travel itself was considered dangerous and experimental. However, as safety improved and air travel became more accessible, public trust grew steadily. A similar evolution is occurring today with electric and autonomous aircraft. The public's perception of safety in aviation has always been tied to visible reliability and consistent regulatory oversight. Electric and hybrid aircraft challenge these perceptions because they use unfamiliar propulsion systems and energy sources. Although electric propulsion reduces mechanical complexity and the risk of engine failure, passengers still express concern about battery reliability, fire hazards, and emergency handling. Overcoming these psychological barriers requires not just technical solutions but also transparent communication and education about safety protocols, performance standards, and environmental benefits.

11.4.2 Factors Influencing Public Perception

Several interrelated factors shape how the public perceives safety and trust in emerging aviation systems. Media coverage plays a significant role in amplifying incidents or highlighting progress. Positive stories about sustainability, cost efficiency, and quiet operation can enhance acceptance, while reports of accidents or malfunctions can quickly erode trust.

Cultural attitudes toward technology also differ across regions. In countries with high technological literacy and strong regulatory institutions, people tend to be more open to innovation. In contrast, in societies with limited exposure to aviation or lower institutional trust, skepticism remains high.

Furthermore, direct experience plays a crucial role. Pilot programs, public demonstrations, and passenger trials allow individuals to see and experience the technology firsthand, reducing fear of the unknown. Public confidence grows when safety records are transparent, and operations are visibly well-regulated.

11.4.3 Safety Concerns in Emerging Air Mobility

Safety is the most critical determinant of public acceptance. While conventional aviation boasts a strong safety record, new electric and autonomous systems introduce unique risks that must be managed effectively. One major concern involves battery safety—particularly thermal runaway events that can cause fires or explosions if not properly controlled. Ensuring robust battery management systems and thermal insulation is vital. Another safety concern relates to autonomous operation. Many next-generation aerial vehicles are designed to operate with limited human intervention. This raises questions about system redundancy, decision-making algorithms, and response in emergency scenarios. The public often associates automation with uncertainty, fearing loss of human control. Building trust in autonomous flight will depend on transparent certification processes and successful real-world testing. Additionally, air traffic integration presents challenges. As drones, air taxis, and electric aircraft share airspace, the risk of mid-air conflicts increases. Developing advanced air traffic management systems, including Unmanned Traffic Management (UTM), is essential to maintain safety and order. Noise perception, privacy concerns, and ground safety—particularly in urban areas—also influence public opinion about the widespread use of electric aircraft and drones.

Table: Major Public and Safety Concerns

Concern Category	Description	Impact on Public Perception
Battery Reliability	Fear of overheating or fire in lithium-based systems	Reduces trust in electric propulsion
Autonomy and Control	Concerns about AI-based decision-making	Creates fear of system failure
Airspace Safety	Potential collisions with manned or unmanned aircraft	Raises regulatory and safety doubts
Noise and Privacy	Worries about noise pollution and camera surveillance	Causes resistance in urban areas
Certification Standards	Uncertainty about global safety regulations	Slows adoption and public approval

11.4.4 Role of Regulation and Certification

Public confidence in aviation has always depended on the credibility of regulatory bodies. Agencies such as the Federal Aviation Administration (FAA), the European Union Aviation Safety Agency (EASA), and the International Civil Aviation Organization (ICAO) play a crucial role in shaping how the public perceives new technologies. Certification standards for electric aircraft, hybrid propulsion systems, and autonomous operations are currently being developed and refined.

Regulators are implementing step-by-step certification procedures to ensure that electric aircraft meet the same safety standards as conventional ones. Transparent evaluation of crashworthiness, battery testing, emergency evacuation, and redundancy in control systems are vital to maintaining credibility. Furthermore, international harmonization of standards ensures that safety protocols are consistent across borders, promoting global trust in electric aviation systems.

Public education programs led by aviation authorities and manufacturers can also help clarify misconceptions. By explaining how safety features like multiple redundant systems, real-time monitoring, and predictive maintenance work, regulators and manufacturers can enhance confidence in the reliability of electric and autonomous flight.

11.4.5 Communication and Media Influence

Communication plays a decisive role in shaping public perception. The way the media presents aviation innovations influences how quickly the public adopts them. Overemphasis on failures, crashes, or technical problems can trigger fear and resistance, while balanced reporting that includes safety improvements and environmental benefits promotes acceptance. Manufacturers and government agencies must actively engage in transparent communication. Public demonstrations, open safety briefings, and digital campaigns showing real-world performance data can help demystify new technologies. When passengers understand how these systems work, they are more likely to trust them. Trust, in turn, directly impacts investment, adoption rates, and regulatory support.

11.4.6 Psychological and Social Dimensions of Safety

Safety is not purely technical—it is also psychological. Even statistically safe technologies can face rejection if people perceive them as risky. Studies on technology acceptance show that emotional factors such as fear, control, and familiarity influence decision-making. Passengers prefer systems they can understand and control, which is why pilot-operated aircraft still inspire greater confidence than autonomous drones. To overcome these barriers, developers must focus on human-centered design, ensuring comfort, reliability, and ease of use. Cockpit visibility, smooth flight experience, and redundant safety alerts all contribute to a sense of control and reassurance. Training programs for pilots, engineers, and operators further strengthen public confidence in the system's reliability.

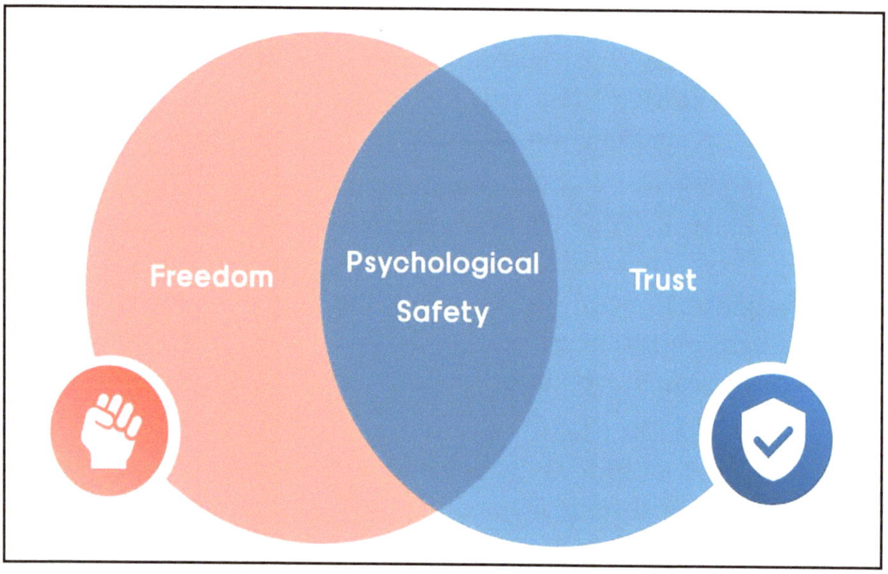

Safety from Psychological and Social Viewpoints

11.4.7 Elements for Enhancing Public Trust and Safety

Clear and transparent regulatory standards for electric and autonomous aircraft.

- Redundant safety systems to prevent single-point failures.
- Rigorous testing of batteries and propulsion units under real conditions.
- Comprehensive pilot and operator training programs.
- Public demonstrations showcasing safe and quiet operation.
- Effective communication about environmental and economic benefits.
- Involvement of local communities in infrastructure planning.
- Collaboration between governments, academia, and industry on safety research.
- Real-time data monitoring and predictive maintenance technologies.
- Integration of human oversight in autonomous flight control systems.

These elements form the foundation for building strong public confidence in next-generation aviation and ensuring safety across all levels of operation.

11.4.8 Case Studies and Public Response

Public perception often evolves as tangible evidence of safety accumulates. For example, Norway's electric flight trials and the USA's urban air mobility pilot programs have shown that exposure to new technology improves trust. Passengers who witness quiet, emission-free flights are more likely to support electric aviation initiatives. Similarly, companies like Joby Aviation and Eviation have emphasized transparent flight testing, live demonstrations, and collaboration with regulators to strengthen credibility.

Community engagement is also vital. When local governments involve residents in decisions regarding vertiport locations or air corridors, resistance decreases. Building trust is a gradual process, and each successful demonstration brings society one step closer to accepting electric and autonomous aircraft as safe, efficient, and reliable.

11.5 AIRSPACE CONGESTION AND INTEGRATION WITH LEGACY SYSTEMS

Airspace congestion and integration with legacy systems represent one of the most complex challenges in modern aviation. As the skies become populated with new types of aircraft—from eVTOLs to autonomous drones—traditional air traffic frameworks must evolve beyond their historical limitations. Achieving this transformation requires a blend of advanced technology, updated regulations, and continued human oversight. By developing integrated systems that combine automation, data-driven management, and sustainable design, the aviation industry can unlock the full potential of urban air mobility while maintaining the safety and efficiency that define modern flight. In the years ahead, the successful coexistence of legacy and future systems will determine not only the efficiency of global transportation networks but also the sustainability and accessibility of aviation in the twenty-first century.

11.5.1 Growing Challenge of Airspace Congestion

Airspace congestion refers to the overcrowding of flight paths and control zones due to increasing air traffic volume. In recent decades, global air traffic has surged due to population growth, urbanization, and economic globalization. The proliferation of commercial aviation, coupled with emerging air mobility technologies, has created a dense and dynamic aerial ecosystem. In urban settings, eVTOL aircraft and delivery drones are expected to operate in lower altitudes traditionally reserved for helicopters and light aircraft. This additional layer of air traffic intensifies the complexity of flight coordination and monitoring, particularly in metropolitan regions with multiple airports and restricted zones.

Congestion poses risks that extend beyond flight delays. It can lead to reduced operational safety margins, increased fuel consumption due to holding patterns, and heightened workload for air traffic controllers. Moreover, as aircraft density increases, the probability of near-miss incidents and communication errors also rises. Without proper management, these challenges can undermine public confidence in both traditional aviation and emerging air mobility systems. The key to resolving these issues lies in redesigning airspace management to accommodate both legacy systems and the diverse requirements of modern aircraft.

11.5.2 Structure and Limitations of Legacy Air Traffic Systems

Legacy air traffic management systems were developed in the mid-20th century to serve a relatively predictable aviation environment dominated by commercial airlines and general aviation. These systems rely heavily on human operators, radio communications, radar surveillance, and centralized control. While highly reliable and safe, they were not designed to handle the large-scale integration of autonomous or semi-autonomous aerial vehicles. The operational model is based on hierarchical control zones, fixed routes, and flight plans submitted in advance—an approach ill-suited for the dynamic and decentralized nature of low-altitude air mobility.

11.5.3 Role of eVTOLs and Urban Air Mobility in Airspace Demand

Electric vertical takeoff and landing aircraft are at the forefront of the next aviation revolution. Designed to transport passengers and cargo across short urban routes, eVTOLs promise to relieve ground congestion and offer sustainable transport solutions. However, their deployment significantly increases low-altitude airspace usage, creating new operational pressures. Unlike traditional aircraft that follow structured flight corridors, eVTOLs require flexible routing capabilities for efficient point-to-point travel.

This shift challenges conventional notions of controlled and uncontrolled airspace, necessitating new frameworks for communication, navigation, and separation assurance. Moreover, eVTOL operations demand integration with existing airport infrastructure and air traffic procedures. As urban air mobility grows, thousands of small-scale flights could occur daily within the same airspace that currently handles helicopters, general aviation, and commercial jets. Coordinating this traffic safely requires advanced systems capable of real-time data exchange between pilots, automated platforms, and control authorities. Such a system must not only prevent conflicts but also ensure efficient airspace utilization without disrupting established flight corridors.

CHAPTER 12

The Future of Electric Aviation

12.1 ROADMAP TO 2030 AND 2050: FORECASTS AND PROJECTIONS

The roadmap to 2030 and 2050 is not a single forecast but a set of conditional projections sensitive to policy, technology and social choices made in the coming decade. Key indicators to watch include battery energy densities and costs, the pace of grid modernization investments, regulatory milestones for autonomy and urban air operations, and patterns of public acceptance. If early pilots are paired with systemic investments in energy and urban infrastructure, the period after 2030 can deliver broad societal benefits from electrified, autonomous mobility. Without that alignment, progress will be slower and unequally distributed. Ultimately, the choices made now determine whether the 2050 horizon is one of integrated, low-carbon mobility or one of persistent fragmentation and missed opportunities.

12.1.1 Drivers Shaping the Next Three Decades

Several powerful drivers determine trajectories to 2030 and 2050: the pace of decarbonization and energy transition, advances in autonomy and digital infrastructure, urbanization patterns, and shifting consumer expectations around convenience and affordability. Battery chemistry and alternative energy carriers influence mission capability for electric mobility and air vehicles, while progress in artificial intelligence and sensing affects the safety envelope for pilotless operations. Policy choices about carbon pricing, urban planning, and public investment in grids and multimodal nodes create enabling or constraining conditions. Financial markets play a gatekeeping role by determining which technologies receive patient capital for long certification and infrastructure cycles, and public acceptance remains a wildcard that can accelerate or stall deployments.

12.1.2 2025–2030: Consolidation, Pilots and Early Scale

In the near term the most plausible scenario through 2030 involves targeted commercialization in constrained markets rather than broad consumer adoption. Urban air mobility and advanced air mobility services will likely operate in demonstration corridors, airport shuttles, and high-value point-to-point routes where demand justifies premium pricing and regulatory frameworks are permissive. On the ground, electric vehicles will continue rapid uptake for light-duty transport, supported by expanding charging networks, while heavy-duty electrification and freight electrification progress more slowly where energy density limits and capital costs are binding. Grid operators will increasingly pilot managed charging and localized storage paired with fast chargers and depots to avoid immediate distribution upgrades, and standards for interoperability and cybersecurity will begin to coalesce as stakeholders learn from early deployments.

12.1.3 2030–2040: Scaling, Integration and Economic Tipping Points

The decade from 2030 to 2040 is the period in which technology maturation and policy alignment can yield tipping points. Battery energy density improvements, cost declines, and modular manufacturing could reduce unit costs and extend the practical range of electric aircraft and long-haul EVs, while more mature autonomy stacks and verified safety cases might permit conditional beyond-visual-line-of-sight and pilot-optional operations in controlled airspaces. If grid modernization and regulatory reforms keep pace, high-capacity charging hubs and vertiports could proliferate, integrated with local storage and renewables to limit peak demand. Economies of scale in manufacturing and operations could push unit costs down, making services more accessible beyond early premium segments, and multimodal trip planning could integrate air, rail, and ground services into seamless chains that change commuting and logistics patterns.

12.1.4 2040–2050: Deep Decarbonization and Normative Transformation

By 2050 the structural outcome will hinge on cumulative investment choices and technological progress achieved earlier. In an ambitious pathway, deep decarbonization of electricity coupled with widespread electrification of transport, high-capacity energy storage deployment, and efficient, integrated mobility systems results in radically lower operational emissions across sectors. Urban air mobility could operate as a mature layer for both passengers and freight, with autonomous fleets providing high-frequency, regulated services that complement ground transit rather than duplicate it. Cities may be reconfigured around multimodal hubs and reduced road demand, yielding changed land use and lower congestion. Conversely, a fragmented pathway with uneven grid upgrades, lagging battery improvements, and social resistance to noise and privacy impacts would limit adoption to niche markets and leave many opportunities unrealized.

12.1.5 Sector Perspectives: Mobility, Energy and Urban Systems

Transport electrification and autonomy interact intimately with energy systems. For road transport, the 2030 horizon will see substantial EV market share in new sales in many regions, but a full fleet turnover requires decades and will depend on scrappage rates, policy incentives, and residual value dynamics. Aviation's transformation is more uneven; small electric and hybrid aircraft, including eVTOLs, may be common in regional and urban contexts by mid-century, while long-haul aviation still depends on sustainable aviation fuels or novel energy carriers unless breakthrough energy densities emerge. Energy systems will need vast investments in transmission, distribution and storage to absorb electrified transport loads; smart charging, vehicle-to-grid capabilities, and decentralized renewables will be essential elements to manage variability and costs while preserving reliability. Urban planning that prioritizes integrated hubs and right-sized infrastructure will enable synergies, but misalignment between transport and energy investment schedules can create bottlenecks.

12.2 INNOVATION IN AIRCRAFT AND BATTERY MATERIALS

12.2.1 Advanced Composites and Lightweight Structures

Aircraft construction is experiencing a sweeping shift away from traditional aluminum and titanium towards high-performance composite materials that offer superior strength-to-weight ratios, exceptional corrosion resistance, and remarkable fatigue durability. Carbon fiber-reinforced polymers (CFRPs) and thermoplastic composites represent the cutting edge, increasingly displacing metals in primary structures such as fuselages, wings, and empennages. CFRPs contribute to significant weight reductions, directly translating into improved fuel performance and extended range for both passenger jets and eVTOL aircraft.

Recent advancements include the development of biomimetic wing structures with increased aerodynamic efficiency and the exploration of recyclable and bio-based matrix systems that further push sustainability goals. Thermoplastic composites, as opposed to thermoset types, greatly streamline manufacturing through their melting and reshaping properties. Automated fiber placement and additive manufacturing enable precise fabrication and geometric flexibility, reducing waste and optimizing material distribution. Ongoing projects integrate composite matrix innovations for engine fan blades and fuselage sections, offering improved efficiency and resilience while supporting complex aerodynamics and structural requirements for next-generation aircraft.

12.2.2 Sustainable and Recyclable Aerospace Materials

Environmental responsibility drives the aerospace sector's material development strategies, with research intensely focused on circular economy principles, recyclability, and minimizing the environmental impact of aircraft production. The industry is increasingly leveraging bio-resins, natural fibers, and recycled composites in manufacturing airframe components, seeking to replace petroleum-derived polymers and foster greater resource sustainability. Standardization efforts, propelled by regulatory and market forces, encourage consistency in manufacturing processes and material qualification, aiming to make composite recycling a scalable reality across all commercial fleets. Durable, low-emission composites reduce lifecycle carbon footprints, and the adoption of sustainable matrix formulations delivers meaningful progress toward meeting international climate goals and regulatory mandates for emission reductions.

12.2.3 Multifunctional and Structural Battery Composites

The intersection of material and energy system innovation is best exemplified in structural battery composites, which integrate energy storage directly into the physical structure of the aircraft. This technology merges advanced composite materials with battery cells to form load-bearing structures that simultaneously store energy, allowing designers to eliminate heavy, redundant battery enclosures and wiring. Such developments not only offer substantial weight savings but also facilitate the design of all-electric and hybrid-electric aircraft with much higher energy density and operational flexibility. Structural batteries are being prototyped for multifunctional fuselage and wing structures, with performance targets set to accelerate certification and commercial deployment in the coming decade. The harmonization of composite material properties with battery performance is driving the sector toward radically more integrated, lighter, and more electrified airframes, supporting longer-range and higher-capacity electric aircraft operations.

12.3 ROLE IN DECARBONIZING GLOBAL TRANSPORTATION

The role of transportation in global decarbonization is both transformative and indispensable. Every mode of travel—from cars and trains to aircraft and ships—must evolve toward cleaner, renewable-powered systems. The aviation industry, once a symbol of carbon intensity, is now leading the charge through electric propulsion, sustainable fuels, and energy-efficient designs. Ultimately, decarbonizing global transportation is a collective endeavor requiring innovation, policy support, and societal participation. As technology advances and public awareness grows, the dream of a carbon-neutral transportation system is becoming an achievable reality—one that will shape the sustainable, connected, and resilient world of tomorrow.

12.3.1 Global Transportation and its Carbon Footprint

Transportation's contribution to global emissions stems primarily from its heavy reliance on fossil fuels. Road vehicles account for the largest share, followed by aviation and maritime transport. Each mode faces distinct challenges in achieving carbon neutrality due to differences in energy density requirements, infrastructure availability, and operational patterns. Aviation remains particularly difficult to decarbonize because jet fuel offers high energy per unit weight—something batteries currently cannot match. Nevertheless, electric and hybrid-electric propulsion systems, sustainable aviation fuels (SAF), and hydrogen-based solutions are being developed to reduce aviation's carbon footprint. The synergy between these technological advances and supportive policies forms the foundation of global decarbonization strategies.

12.3.2 Transition Pathways for Decarbonization

The path to a decarbonized transportation network is multifaceted and region-specific. Developed economies such as those in the EU, USA, and Japan focus on large-scale electrification, renewable energy integration, and green hydrogen development. Developing nations, on the other hand, prioritize affordable clean energy access and infrastructure modernization. The aviation sector exemplifies this transition through initiatives like hybrid-electric regional aircraft, all-electric short-haul planes, and the use of SAF derived from waste oils, biomass, and synthetic sources. Maritime industries are experimenting with low-carbon fuels and wind-assist systems, while urban transportation systems are adopting electric buses and rail networks powered by renewable electricity. Governments are also implementing carbon pricing mechanisms, incentives, and emissions trading systems to promote sustainable transport investments. The combination of innovation, regulation, and behavioral change will be the key to achieving meaningful carbon reductions across all modes of transport.

12.3.3 Aviation Sector's Contribution to Decarbonization

Aviation contributes approximately 2–3% of total global CO_2 emissions, but its impact is amplified by the high-altitude release of greenhouse gases. Electric aircraft, hybrid propulsion systems, and SAFs are among the leading solutions being tested to decarbonize air travel. Electric aviation, in particular, holds promise for short-haul routes and regional connectivity. Companies such as Eviation, Heart Aerospace, and Pipistrel are developing aircraft that run entirely or partially on electric power, significantly cutting emissions and operating costs. Hybrid-electric propulsion, which combines a conventional engine with electric systems, can reduce fuel consumption by up to 30–40%. SAFs can cut lifecycle emissions by up to 80%, providing an immediate pathway toward lower-carbon flight without requiring major infrastructure changes. Over time, integrating these technologies with renewable energy systems will enable a fully carbon-neutral aviation ecosystem.

12.3.4 Interconnection between Transportation Modes

Decarbonization cannot occur in isolation—intermodal coordination is essential. For example, electric aviation must rely on renewable electricity generated from solar, wind, or hydro sources to maintain net-zero benefits. Similarly, airports, ports, and urban transport hubs need integrated energy infrastructure to support electric charging and hydrogen refueling. The convergence of technologies such as smart grids, electric vehicle charging stations, and renewable microgrids ensures energy efficiency across transport networks. The electrification of aviation can thus complement the broader transition to clean energy, creating a feedback loop that accelerates sustainability across multiple modes of travel.

12.3.5 Strategies for Decarbonizing Global Transportation

- Expanding renewable energy production to power electric vehicles and aircraft.
- Investing in sustainable aviation fuels to replace conventional jet fuels.
- Electrifying urban public transport systems to reduce emissions in cities.
- Encouraging modal shifts toward rail and sea for long-distance freight.
- Introducing global carbon pricing and emissions reduction targets.
- Supporting innovation in hydrogen and ammonia fuel technologies.
- Developing recycling systems for batteries and green energy components.
- Enhancing urban planning to reduce commuting distances and congestion.
- Creating international standards for carbon accounting and reporting.
- Promoting public awareness and behavioral changes toward sustainable travel.

These strategies emphasize the collaborative effort required from governments, industries, and individuals to create a low-carbon transportation ecosystem that supports both mobility and climate resilience.

12.3.6 Policy Frameworks and Global Initiatives

Policy initiatives play an indispensable role in accelerating decarbonization. The European Green Deal, the U.S. Inflation Reduction Act, and China's Dual Carbon Goals have set ambitious targets for carbon neutrality by mid-century. These policies promote clean energy investments, subsidize electric vehicle adoption, and fund green infrastructure. In the aviation sector, the Carbon Offsetting and Reduction Scheme for International Aviation (CORSIA), developed by ICAO, mandates carbon-neutral growth beyond 2020 by offsetting emissions through verified carbon credits. Meanwhile, the Clean Skies for Tomorrow initiative encourages collaboration among airlines, airports, and manufacturers to scale SAF production. In India, initiatives like the National Electric Mobility Mission Plan (NEMMP) and FAME (Faster Adoption and Manufacturing of Electric Vehicles) are supporting e-mobility adoption across sectors. The collective implementation of such frameworks demonstrates how international coordination can drive real progress toward decarbonization.

12.3.7 Technological Innovations in Low-Carbon Transport

Technology is the backbone of decarbonization. In aviation, innovations in electric propulsion, lightweight materials, and high-density batteries are reducing energy consumption and improving range efficiency. Hydrogen-powered aircraft, still in experimental stages, could become viable by the 2030s, offering zero direct emissions and long-range capabilities. Similarly, maritime industries are adopting hybrid-electric propulsion and exploring green ammonia as an alternative to fossil fuels. Road transport is witnessing a revolution through solid-state batteries, which promise faster charging and higher energy storage. In the rail sector, hydrogen fuel cell locomotives and electrified rail corridors are transforming freight and passenger mobility. Digitalization also plays a major role. Smart transport systems use artificial intelligence, predictive analytics, and Internet of Things (IoT) technologies to optimize routes, manage energy consumption, and reduce idle emissions. These innovations collectively make global transportation more efficient, cleaner, and cost-effective.

12.3.8 Economic and Social Implications

Decarbonizing transportation brings not only environmental benefits but also economic and social advantages. Transitioning to electric and hydrogen technologies can create millions of new jobs in renewable energy, manufacturing, and infrastructure development. Cleaner transportation reduces air pollution, leading to improved public health and lower healthcare costs. However, challenges remain, particularly in terms of the initial investment required for infrastructure and the affordability of new technologies. Governments must balance innovation incentives with equitable access to ensure that the benefits of decarbonization are shared across societies. Public-private partnerships are therefore crucial in bridging the financial and technological gaps.

12.3.9 Challenges and Barriers

Despite significant progress, several challenges hinder rapid decarbonization. The limited energy density of current batteries restricts electric aviation to short-range flights. Hydrogen requires extensive infrastructure for production, storage, and transport. Sustainable aviation fuels, while promising, remain expensive and limited in availability. Policy alignment is another barrier—disparate regulatory frameworks across regions can slow progress. Additionally, consumer hesitancy, driven by concerns over range, cost, and safety, affects adoption rates. Addressing these issues requires continued investment in research, international collaboration, and transparent communication about the benefits and reliability of green technologies.

12.4 MILITARY AND SPACE APPLICATIONS OF ELECTRIC PROPULSION

Electric propulsion represents a defining technological advancement with profound implications for both military strategy and space exploration. Its efficiency, adaptability, and potential for silent, long-duration operation make it an ideal solution for modern defense platforms and spacecraft. While challenges related to power generation, cost, and thrust output persist, ongoing research and international collaboration are steadily overcoming these barriers. As innovation accelerates, electric propulsion will not only enhance operational capabilities but also redefine the boundaries of what is possible in air, sea, and space. Ultimately, the convergence of electric power and propulsion marks the beginning of a new era—one where sustainability, precision, and technological mastery drive humanity's journey from Earth's battlefields to the farthest reaches of the cosmos.

12.4.1 Principles and Advantages of Electric Propulsion

Electric propulsion operates on the fundamental principle of converting electrical energy into kinetic energy to produce thrust. Depending on the application, this energy may be derived from onboard batteries, solar panels, or nuclear reactors. Unlike chemical engines that expend fuel explosively, electric propulsion systems release ions or plasma at high velocities, achieving far greater specific impulse—a measure of propulsion efficiency. In practical terms, electric propulsion allows vehicles to travel farther using significantly less propellant, an advantage that has profound implications for both space exploration and military operations. The benefits of electric propulsion extend beyond fuel efficiency. These systems are quieter and generate less heat, offering stealth advantages in military environments. They also reduce logistical demands associated with fuel transport and storage, enhancing the strategic mobility of defense forces. In space missions, the precision and controllability of electric thrusters make them ideal for fine orbital adjustments, long-duration missions, and deep-space exploration. Together, these advantages position electric propulsion as a cornerstone technology for the future of defense and aerospace innovation.

12.4.2 Military Applications of Electric Propulsion

In the military domain, electric propulsion has gained prominence for its potential to revolutionize air, land, and sea operations. Modern defense systems increasingly prioritize energy efficiency, stealth, and endurance—all attributes supported by electric propulsion technologies. Electrically driven systems reduce the acoustic and thermal signatures of vehicles, making them less detectable to enemy sensors. This capability is particularly valuable in submarines, unmanned aerial vehicles (UAVs), and reconnaissance platforms where stealth is critical. For example, electric propulsion systems in naval vessels, such as integrated full electric propulsion (IFEP), allow ships to operate quietly and with greater fuel efficiency. The British Royal Navy's Type 45 destroyers and the U.S. Navy's Zumwalt-class destroyers employ electric propulsion architectures to power both propulsion and onboard systems from a unified electrical grid. This integration enhances operational flexibility and allows future adaptation for directed-energy weapons and advanced radar systems. Similarly, in aviation, the introduction of hybrid-electric propulsion in drones and tactical aircraft offers reduced fuel consumption and prolonged loiter times, extending surveillance and strike capabilities without compromising performance.

Ground-based military vehicles are also transitioning toward hybrid-electric or fully electric systems. Armored vehicles equipped with electric drivetrains benefit from reduced maintenance, enhanced torque, and lower acoustic emissions, which improve battlefield survivability. Moreover, electric propulsion aligns with military goals for logistical sustainability by reducing reliance on fossil fuels, thus minimizing supply chain vulnerabilities in conflict zones. These advancements collectively represent a strategic shift toward more agile, efficient, and resilient defense forces capable of operating effectively across diverse environments.

12.4.3 Space Applications of Electric Propulsion

In space exploration and satellite technology, electric propulsion has become an essential component of modern spacecraft design. Traditional chemical propulsion systems, while powerful, consume vast amounts of propellant for relatively short bursts of thrust. In contrast, electric propulsion offers continuous, gentle acceleration over extended periods, enabling spacecraft to reach higher velocities and travel greater distances with minimal fuel mass. This advantage is especially critical for deep-space missions, where resupply is impossible and efficiency determines mission success.

The most widely used electric propulsion systems in space include ion thrusters, Hall-effect thrusters, and arcjet engines. Ion thrusters, such as those used on NASA's Deep Space 1 and Dawn missions, accelerate ions through electric fields to generate thrust. They have demonstrated exceptional endurance and efficiency, allowing spacecraft to orbit multiple celestial bodies within a single mission. Hall-effect thrusters, favored for commercial satellites, provide a balance between efficiency and thrust, making them suitable for maintaining geostationary orbits and conducting orbital transfers. Arcjet and pulsed plasma thrusters, while less efficient, offer higher thrust-to-power ratios suitable for spacecraft requiring responsive maneuvering capabilities.

Electric propulsion also plays a pivotal role in satellite constellations used for communications, navigation, and Earth observation. With hundreds or even thousands of satellites operating simultaneously, maintaining precise orbital alignment is critical. Electric thrusters provide the continuous fine-tuning necessary for station-keeping and collision avoidance. Furthermore, as space becomes increasingly crowded, electric propulsion allows satellites to perform controlled de-orbiting maneuvers at the end of their service life, supporting international efforts to reduce space debris and promote sustainable orbital practices.

12.4.4 Strategic and Operational Advantages

The adoption of electric propulsion brings multiple strategic advantages that extend beyond performance metrics. In military contexts, the shift toward electrified platforms enhances operational flexibility by simplifying energy distribution across systems. A fully electric vessel or aircraft can allocate power dynamically between propulsion, weapons, and communications, optimizing performance based on mission needs. This adaptability supports emerging technologies such as railguns, laser weapons, and electromagnetic launch systems, all of which require massive electrical power but benefit from integrated power management.

In space exploration, electric propulsion reduces mission costs by lowering fuel requirements and launch mass. This efficiency enables smaller rockets to deliver spacecraft into orbit, democratizing access to space for both government and private organizations. Electric propulsion also supports the growing trend of reusable spacecraft, as thrusters can be refueled or recharged for multiple missions. On interplanetary missions, electric propulsion systems enable sustained acceleration, significantly shortening travel times to distant destinations such as Mars or the asteroid belt. The enhanced precision of electric thrusters allows spacecraft to execute delicate maneuvers for planetary orbit insertion, docking, and resource prospecting—capabilities that are essential for future human exploration missions.

12.4.5 Technological Challenges and Development Barriers

Despite its promise, electric propulsion faces several technological and operational barriers that must be overcome to realize its full potential. One major challenge is power generation. In military applications, batteries and onboard generators must provide sufficient energy to sustain propulsion and auxiliary systems without adding excessive weight. In space, solar arrays and nuclear power sources must deliver continuous, reliable energy in extreme environments where sunlight and temperature vary dramatically. Developing compact, lightweight, and high-output power systems remains a primary focus of research. Another limitation is the relatively low thrust output of electric engines compared to chemical propulsion. While electric systems excel in efficiency, they cannot yet provide the rapid acceleration required for atmospheric flight or heavy payload launches. As a result, hybrid systems that combine electric and chemical propulsion are being developed to bridge this performance gap. Additionally, materials and thermal management pose challenges, as electric thrusters operate at high voltages and temperatures that can degrade components over time. Advances in superconductors, heat-resistant alloys, and plasma containment technologies are essential to improving reliability and scalability.

12.5 GLOBAL COLLABORATION AND POLICY RECOMMENDATIONS

Global collaboration is not a luxury but a prerequisite to realizing safe, equitable and sustainable mobility transformations. Harmonized standards, shared data, and coordinated investment reduce friction, spread innovation benefits and enhance resilience in the face of technological and geopolitical uncertainty. By aligning regulatory ambition with pragmatic governance, financing and community engagement, international actors can ensure that the transition to electrified, autonomous mobility supports climate goals, economic development and the protection of public interest. The choices made today about cooperation frameworks and policy design will determine whether the coming mobility revolution becomes an inclusive global success or a fragmented set of local experiments.

12.5.1 International Cooperation is Essential

Air transport has always been an international enterprise where mutual recognition of standards, interoperable navigation and communication systems, and coordinated traffic management enable efficient cross-border operations. The advent of distributed low-altitude operations and autonomous systems expands the scope of interoperability requirements to include digital platforms, cybersecurity protocols, and real-time data exchange. Without coordinated rules and technical baselines, manufacturers face a patchwork of certification demands, operators encounter regulatory barriers to scale, and urban planners cannot design interoperable vertiports or charging hubs. International cooperation reduces market friction, aligns safety expectations, and creates economies of scale in production and training while enabling developing countries to participate through capacity building and technology transfer.

12.5.2 Regulatory Harmonization and Mutual Recognition

Regulatory frameworks must evolve from prescriptive, jurisdiction-specific mandates to performance-based, risk-proportionate systems that permit innovation while ensuring safety. Harmonization efforts should prioritize mutual recognition agreements that allow certificates, test results, and safety cases developed in one jurisdiction to be accepted elsewhere with proportionate oversight. International organizations such as ICAO, regional regulators and industry standard bodies should collaborate to define common airworthiness principles for electric propulsion, autonomy assurance frameworks for AI-enabled flight systems, and interoperable rules for low-altitude traffic management. Creating a clear, staged certification pathway helps manufacturers and operators plan investments and reduces regulatory latency, while sandbox environments and conditional authorizations permit controlled operational learning without compromising public safety.

12.5.3 Technical Standards, Airspace Management and Data Sharing

Interoperability of technical systems is central to a scalable ecosystem. Standards for communications, navigation and surveillance must enable secure, low-latency links and support resilient alternatives to GNSS where accuracy can be degraded. Traffic management concepts for low-altitude operations require common data models, open APIs, and agreed-upon semantics for flight authorization, dynamic rerouting, and contingency coordination. A neutral, global approach to data governance should balance operational safety needs with privacy protections, enabling anonymized operational datasets to be shared for safety analysis and research. International coordination on radio spectrum allocation and the role of telecommunications providers in supporting command-and-control links will be critical to avoid harmful interference and ensure equitable access to essential resources.

12.5.4 Safety Assurance, Cybersecurity and Liability Frameworks

Safety and security are twin pillars that require harmonized approaches. Certification regimes must incorporate robust methods for validating learning-based components, runtime monitoring, and formalized fallback behaviors. Cybersecurity standards should be embedded into design, supply chain practices and operational procedures, with clear expectations for cryptographic protections, intrusion detection, and incident reporting. Liability frameworks need clarity so that operators, manufacturers, software providers and infrastructure owners understand responsibility in routine and failure scenarios. International coordination on incident investigation processes and a global safety reporting system will accelerate collective learning from accidents and near-misses while preserving legal certainty for commercial deployment.

CHAPTER 13

Appendices

13.1 GLOSSARY OF KEY TERMS

All-Electric Propulsion: A propulsion method in which the aircraft is powered exclusively by electric motors, drawing energy from onboard batteries or other electrical storage systems.

Battery Energy Density: The amount of energy stored per unit mass or volume in a battery. Higher energy density enables longer flight times and greater range for electric aircraft.

Certification and Regulatory Compliance: Standards and regulations established by aviation authorities, such as the FAA or EASA, for the safe design, operation, and maintenance of electric aircraft and associated systems.

Charge/Recharge Infrastructure: Facilities and technologies enabling the safe, rapid charging or swapping of batteries for electric aircraft, critical for operational viability.

Distributed Electric Propulsion (DEP): A configuration using multiple small electric motors and propellers distributed across the airframe, improving lift, efficiency, and redundancy while enabling quieter operation.

Electric Aircraft (eAircraft): An aircraft powered fully or partially by electricity through batteries, fuel cells, or hybrid-electric systems, rather than conventional fossil-fuel engines.

Electric Taxing: The use of electric motors to move aircraft on the ground without running the main engines, reducing fuel consumption and emissions at airports.

Emission Flight: A flight operation where the aircraft produces no harmful emissions, typically achieved through all-electric or hydrogen fuel cell propulsion systems.

Energy Management System (EMS): A control system that optimizes power usage from batteries, fuel cells, or hybrid systems, ensuring efficiency, safety, and performance during all flight phases.

Fuel Cell: A device that converts chemical energy (often from hydrogen) directly into electricity to power electric motors, producing water as the primary byproduct. Fuel cells are used in some electric and hybrid aircraft for emission-free operation.

Hybrid-Electric Propulsion: A propulsion system that combines traditional combustion engines with electric motors, allowing for reduced fuel consumption, lower emissions, and enhanced efficiency.

Lithium-Ion Battery (Li-ion): A widely used type of rechargeable battery in electric aviation known for high energy density, efficiency, and longevity. Li-ion batteries are the standard for many electric aircraft prototypes.

Noise Footprint: The total acoustic impact of an aircraft on the environment, including takeoff, landing, and cruising. Electric propulsion systems are generally quieter, reducing community noise exposure.

Power-to-Weight Ratio: A measure of an aircraft motor or propulsion system's output relative to its weight, critical in electric aviation where batteries can add significant mass.

Range Extender: An auxiliary system, often a small combustion engine or fuel cell, that generates electricity to supplement batteries and increase flight range for hybrid-electric aircraft.

Regenerative Braking: A method by which kinetic energy from slowing the aircraft is converted back into electrical energy and stored in batteries, improving overall energy efficiency.

Thermal Management: Systems and methods designed to regulate temperature in batteries, motors, and electronics to maintain performance and safety in electric aircraft.

Turboelectric Propulsion: A hybrid-electric system where a gas turbine generates electricity that powers electric motors connected to propellers, combining high efficiency with electric distribution flexibility.

Urban Air Mobility (UAM): The concept of using electric aircraft, particularly eVTOLs, for transporting passengers or cargo in urban environments, aimed at reducing congestion and travel times.

Vertical Takeoff and Landing (VTOL): Aircraft capable of taking off and landing vertically, often using electric motors and ducted fans or rotors. Electric VTOL (eVTOL) aircraft are central to urban air mobility solutions.

13.2 TECHNICAL STANDARDS AND CERTIFICATION DOCUMENTS

13.2.1 ASTM F3239-22a—Aircraft Electric Propulsion Systems

ASTM F3239-22a is a crucial standard that outlines airworthiness requirements for the design and installation of electric propulsion systems in aircraft. It implicitly covers hybrid-electric systems unless explicitly stated otherwise. The specification emphasizes conventional system layouts, propulsion characteristics, and operations. This standard is recognized by aviation authorities such as the FAA and EASA as an Accepted Means of Compliance (AMC) for electric propulsion systems in aircraft.

13.2.2 EASA Certification Specifications (CSs) and Guidance

The European Union Aviation Safety Agency (EASA) provides a comprehensive set of Certification Specifications (CSs) and Acceptable Means of Compliance (AMCs) for various aircraft types, including those with electric propulsion. EASA's documents cover areas such as:

- Electrical system descriptions and load analysis.
- Flight test programs.
- Weight and balance calculations.
- Wing load assessments.
- Minor change certification.

13.2.3 ISO/TC 20/SC 1—Aerospace Electrical Requirements

The International Organization for Standardization (ISO) Technical Committee 20, Subcommittee 1 (TC 20/SC 1), focuses on aerospace electrical requirements. This subcommittee develops standards related to:

- Generation, control, and distribution of electrical energy for civil aircraft.
- Aircraft electric cables.
- Aircraft circuit breakers.
- Solid-state remote power controllers.
- LED power lighting.

13.2.4 US FAA and EASA Harmonization

The Federal Aviation Administration (FAA) and EASA have established Technical Implementation Procedures (TIP) to harmonize certification processes. These procedures facilitate mutual recognition of design approvals, airworthiness, and environmental requirements, ensuring that aircraft certified in one jurisdiction are accepted in the other. This collaboration is particularly beneficial for manufacturers of electric aircraft seeking international certification.

13.2.5 Additional Resources

- **Purdue Polytechnic:** Offers educational materials on airworthiness requirements for electric propulsion systems, including ASTM F3239-22a.

- **EASA Certification Memoranda:** Provide guidance on specific certification items, including those related to electric propulsion.

13.3 ELECTRIC AIRCRAFT COMPARISON TABLES

13.3.1 Electric Aircraft Comparison Table

Aircraft Model	Type	Passengers	Range (km)	Max Payload (kg)	Power Source	Notes
9Bolt	Fully Electric	9	~300	~900	Battery-electric	Designed for short regional routes; low emissions.
19Bolt	Fully Electric	19	~500	~1,700	Battery-electric	Suitable for regional commuter services; reduced operational costs.
90Bolt	Fully Electric	90	~800	~3,000	Battery-electric	Targets regional airliner market; significant CO_2 reduction potential.
E19	Hybrid-Electric	19	~600	~1,800	Hybrid (battery + fuel)	Combines electric propulsion with fuel engine; extended range.
Electra EL9	Hybrid-Electric	9	~500	~1,360	Hybrid-electric	Features ultra-short takeoff and landing capabilities; versatile operations.

13.3.2 Key Performance Metrics

- **Energy Efficiency:** Electric motors typically offer 2–3 times higher energy efficiency in cruise mode compared to conventional fossil-fuel engines.
- **CO_2 Emissions Reduction:** Fully electric aircraft can achieve up to 88% reduction in CO_2 emissions compared to traditional aircraft, depending on the energy mix used for charging.
- **Battery Limitations:** Current battery technologies constrain the range and payload of electric aircraft, with energy densities significantly lower than that of jet fuel.

13.3.3 Visual Payload-Range Comparison

For a visual representation of how payload affects range in electric aircraft, refer to the following diagram:

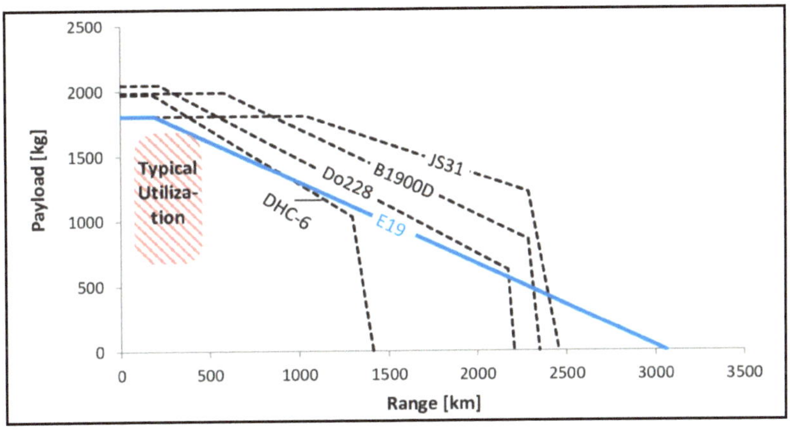

13.4 FUNDING AGENCIES AND RESEARCH PROGRAMS

13.4.1 US United States

ARPA-E (Advanced Research Projects Agency-Energy)

ARPA-E, part of the U.S. Department of Energy, funds innovative projects aimed at transforming the energy sector. Notably, the REEACH program supports advancements in electric aviation by funding projects focused on electric propulsion systems and energy storage solutions. Additionally, the ASCEND program promotes the development of sustainable aviation technologies, including electric and hybrid-electric propulsion systems. These initiatives collectively received $33 million in funding for 17 projects.

FAA Aviation Research Grants Program

The Federal Aviation Administration (FAA) offers the Aviation Research Grants Program, which anticipates $6 million annually for new and continuing grant awards. This program funds independent research on various topics related to the long-term growth of civil aviation, including electric aircraft technologies.

U.S. Air Force SBIR/STTR Program

The Department of the Air Force's Small Business Innovation Research (SBIR) and Small Business Technology Transfer (STTR) programs support the development of innovative technologies. ZeroAvia, for instance, received a grant to explore the integration of autonomy and fuel cells in hydrogen-electric aircraft, enhancing national defense capabilities.

13.4.2 EU Europe

Clean Sky Joint Undertaking

Clean Sky is a public-private partnership between the European Commission and the European aeronautics industry. It coordinates and funds research activities to deliver significantly quieter and more environmentally friendly aircraft. The program focuses on various aspects of electric aviation, including propulsion systems and aircraft design.

UK Government Initiatives

The UK government has announced a £250 million investment in clean technology aimed at making air travel more environmentally friendly. This initiative supports research and development in zero-emission flying, including electric aircraft technologies. Key beneficiaries include major players like Airbus and Rolls-Royce, alongside smaller companies and academic institutions.

Dovetail Electric Aviation

Dovetail Electric Aviation, in collaboration with Aciturri and ITP Aero, secured funding through Spain's Science and Innovation Missions Program. This support facilitates the development of the first hydrogen-electric conversion of a CS-23 aircraft in the EU, focusing on retrofit-ready zero-emission aircraft.

13.4.3 Global Research Institutions

NASA (National Aeronautics and Space Administration)

NASA's Glenn Research Center leads the Electrified Aircraft Propulsion program, focusing on the development of new aviation technologies to enable more efficient commercial air transportation. The program encompasses various aspects of electric aviation, including propulsion systems and aircraft design.

German Aerospace Center (DLR)

The German Aerospace Center (DLR) is involved in numerous research and development projects in the field of electric aviation. As Germany's national center for aerospace, energy, and transportation research, DLR collaborates on projects aimed at advancing electric and hybrid-electric aircraft technologies.

13.4.4 Academic Institutions

Purdue Polytechnic Institute

Purdue Polytechnic Institute received a nearly $100,000 grant from the National Institute of Standards and Technology (NIST) to develop educational modules focused on the safety and design standards for electric aircraft propulsion. This initiative aims to enhance academic curricula and research in electric aviation.

University of Connecticut (UConn)

A research team at UConn, supported by a $4.5 million cooperative agreement from ARPA-E, is developing a high-performance metal-supported solid oxide fuel cell system for range extension of commercial aviation. This project focuses on creating carbon-neutral energy storage and power generation systems for electric aircraft.

13.5 FURTHER READING AND RESOURCE LINKS

13.5.1 Foundational Texts & Technical Books

- **Fundamentals of Electric Aircraft, Revised Edition (SAE R-563):** A comprehensive guide detailing the principles of electric aircraft, including design considerations, propulsion systems, and energy storage solutions.
- **Advances in Electric Aviation:** A compilation of proceedings from the International Symposium on Electric Aviation and Autonomous Systems, covering topics like electrification of aerial vehicles and energy storage.
- **Emerging Trends in Electric Aviation:** Insights from the 2022 symposium, discussing hybrid, electric, and all-electric propulsion systems, as well as fuel alternatives.

13.5.2 Research Papers & Technical Reports

- **Electric Aviation: A Review of Concepts and Enabling Technologies:** An in-depth review identifying key technological areas such as battery technology, electric machines, and propulsion systems.
- **Powering the Future of Electric Aviation:** A NASA report examining the potential of electric aviation in reducing aviation's impact on climate change, highlighting challenges like battery energy density and charging infrastructure.
- **Feasibility Study of an All-Electric 150 Passenger Aircraft:** A NASA study assessing the viability of large-scale electric aircraft, concluding that current battery technologies are insufficient for such applications.

13.5.3 Institutional & Government Resources

- **NASA Electrified Aircraft Propulsion (EAP) Program:** NASA's EAP initiative focuses on improving efficiency and reducing energy consumption in aviation through innovative technologies and flight demonstrations.
- **Electrification of Aircraft: Challenges, Barriers, and Potential:** A report by the National Renewable Energy Laboratory discussing the challenges and potential of aircraft electrification, including propulsion system considerations.

- **Aviation Research Grants Program (U.S. Department of Transportation):** A program funding independent research on various topics related to the long-term growth of civil aviation, including electric aircraft technologies.

13.5.4 Additional Resources

- **Purdue Polytechnic Institute—Electric Aircraft Propulsion:** Educational materials on airworthiness requirements for electric propulsion systems.
- **Dovetail Electric Aviation—Zero-Emission Aircraft Development:** Information on Spain's Science and Innovation Missions Program supporting the development of retrofit-ready zero-emission aircraft.

References

Advanced Battery Systems for Aviation, Author: Dr. Kevin Zhang, Publisher: Springer, 2022

Advanced Capacitor Technology for Aviation, Author: Dr. Rachel Green, Publisher: Elsevier Science, 2021

Aerodynamics of Electric Aircraft Design, Author: Dr. Francesco Rossi, Publisher: AIAA, 2019

Artificial Intelligence in Electric Aviation, Author: Dr. Priya Sharma, Publisher: MIT Press, 2021

Avionics for Electric Aircraft, Author: Captain Laura Bennett, Publisher: Aviation Technology Publishers, 2021

Battery Management Systems for Flight, Author: Dr. Linda Morrison, Publisher: Artech House, 2021

Battery Technology for Electric Aviation, Author: Sarah Chen & Robert Williams, Publisher: Springer Nature, 2021

Brushless DC Motors in Aircraft, Author: Dr. Hiroshi Yamamoto, Publisher: IEEE Press, 2019

Composite Materials for Electric Planes, Author: Dr. Richard Evans, Publisher: Woodhead Publishing, 2019

Convertible Electric Aircraft Design, Author: Dr. Stephanie Morris, Publisher: Wiley, 2021

Cryogenic Systems for Electric Propulsion, Author: Dr. Ivan Sokolov, Publisher: Academic Press, 2020

Design Principles of Electric Airplanes, Author: Andreas Mueller, Publisher: SAE International, 2021

Distributed Electric Propulsion Aircraft, Author: Dr. Lisa Freeman, Publisher: MIT Press, 2021

Electric Aircraft Acoustic Engineering, Author: Dr. Christine Dubois, Publisher: Springer, 2020

Electric Aircraft Charging Infrastructure, Author: Mark Davidson, Publisher: CRC Press, 2022

Electric Aircraft Cooling Solutions, Author: Dr. Victor Popov, Publisher: Springer, 2020

Electric Aircraft Fire Safety, Author: Fire Safety Institute, Publisher: NFPA (National Fire Protection Association), 2021

Electric Aircraft Fundamentals, Author: James Morrison, Publisher: Wiley-IEEE Press, 2019

Electric Aircraft Future Trends, Author: Dr. Victoria Armstrong, Publisher: Springer Nature, 2023

Electric Aviation Weather Systems, Author: Dr. Douglas Mitchell, Publisher: American Meteorological Society, 2022

Electric Aviation Workforce Development, Author: Education Task Force, Publisher: McGraw-Hill Education, 2022

Electric Aviation: From Theory to Practice, Author: Marco Bianchi, Publisher: John Wiley & Sons, 2019

Electric Flight Control Systems, Author: Commander David Clark, Publisher: McGraw-Hill, 2020

Electric Flight: The Zero-Emission Future, Author: Daniel Roberts, Publisher: Routledge, 2022

Electric Glider Technology, Author: Hermann Kruger, Publisher: Soaring Society Publications, 2018

Electric Helicopter Technology, Author: Captain Steven Miller, Publisher: Helicopter Association International, 2021

Electric Motors in Aerospace Applications, Author: Professor David Kumar, Publisher: Elsevier, 2018

Electric Propeller Design and Analysis, Author: Dr. Samuel Johnson, Publisher: AIAA, 2019

Electric Propulsion for Drones, Author: Dr. Omar Abdullah, Publisher: Elsevier, 2019

Electric Propulsion Systems Engineering, Author: Dr. Jennifer Martinez, Publisher: AIAA (American Institute of Aeronautics and Astronautics), 2020

Electric Regional Aircraft Development, Author: Dr. Graham Scott, Publisher: Cambridge University Press, 2021

Electric Unmanned Aerial Vehicles, Author: Colonel Robert Jackson, Publisher: McGraw-Hill, 2018

Electric Vertical Takeoff and Landing Systems, Author: Dr. Patricia O'Brien, Publisher: Cambridge University Press, 2023

Electromagnetic Compatibility in Electric Aircraft, Author: Dr. Klaus Weber, Publisher: IEEE Press, 2020

Energy Management in Electric Aircraft, Author: Professor Hans Schmidt, Publisher: Springer International, 2020

Flight Testing Electric Aircraft, Author: Test Pilot Association, Publisher: AIAA, 2020

Fuel Cell Technology for Aircraft, Author: Professor John Hamilton, Publisher: Wiley-Blackwell, 2020

High-Voltage Systems in Aircraft, Author: Engineering Team - Boeing Research, Publisher: IEEE Press, 2019

Hybrid Turbine-Electric Architecture, Author: Dr. Jonathan Wright, Publisher: AIAA, 2021

Hybrid-Electric Propulsion for Aircraft, Author: Dr. Michael Patterson, Publisher: Cambridge University Press, 2020

Hydrogen Fuel Cells for Electric Flight, Author: Dr. Gabriela Schneider, Publisher: Elsevier, 2022

Lithium-Ion Batteries in Flight Applications, Author: Dr. Sophia Lee, Publisher: Springer, 2021

Magnetic Gearboxes in Aviation, Author: Dr. Robert Fischer, Publisher: IEEE Press, 2020

Modular Electric Propulsion Units, Author: Dr. Antonio Silva, Publisher: Springer, 2022

Motor Control Strategies for Flight, Author: Professor Michael Kim, Publisher: Elsevier, 2020

Multi-Rotor Electric Aircraft Design, Author: Dr. Raymond Chen, Publisher: SAE International, 2020

Next Generation Electric Propulsion, Author: Dr. Wei Liu, Publisher: Springer Nature, 2022

Power Conversion Systems for Flight, Author: Dr. Maria Garcia, Publisher: Elsevier, 2020

Power Density in Aviation Batteries, Author: Dr. Catherine Park, Publisher: Springer, 2022

Power Electronics for Electric Aircraft, Author: Richard Thompson, Publisher: CRC Press, 2019

Power Grid Integration for Electric Airports, Author: Dr. Edward Thompson, Publisher: CRC Press, 2022

Range Optimization in Electric Flight, Author: Dr. Carlos Mendoza, Publisher: Springer, 2020

Regenerative Systems in Electric Flight, Author: Dr. Margaret Sullivan, Publisher: Academic Press, 2021

Small Electric Aircraft Design, Author: Paul Henderson, Publisher: Aviation Publishers, 2019

Solar-Electric Aircraft Systems, Author: Dr. Pierre Dubois, Publisher: Wiley, 2019

Solid-State Batteries for Aircraft, Author: Dr. Elizabeth Turner, Publisher: Springer Nature, 2023

Superconducting Motors for Aviation, Author: Dr. Alexander Petrov, Publisher: Academic Press, 2020

Supersonic Electric Propulsion Concepts, Author: Dr. Alan Richardson, Publisher: AIAA, 2023

Sustainable Aviation through Electrification, Author: Dr. Emily Richardson, Publisher: Academic Press, 2020

Thermal Management for Electric Aviation, Author: Dr. Yuki Tanaka, Publisher: Elsevier Science, 2021

Urban Air Mobility: Electric Flight Revolution, Author: Thomas Anderson, Publisher: McGraw-Hill Education, 2022

Urban eVTOL Operations Manual, Author: Dr. Jessica Rivera, Publisher: Routledge, 2022

Variable Pitch Propellers for Electric Aircraft, Author: Dr. Lars Andersson, Publisher: Cambridge University Press, 2020

Wireless Power Transfer for Aviation, Author: Professor Timothy Brown, Publisher: Cambridge University Press, 2021

Index

www.ingramcontent.com/pod-product-compliance
Lightning Source LLC
LaVergne TN
LVHW071245210926
848486LV00008B/349